SO-BFD-765

Understanding Telephone Electronics

Written by: John L. Fike, Ph.D., P.E.
Adj. Assoc. Professor of Electrical Engineering
Southern Methodist University
Staff Consultant, Texas Instruments Learning Center

George E. Friend
Consultant, Telecommunications
Dallas, Texas
Staff Consultant, Texas Instruments Learning Center

With contributions by: *Texas Instruments Learning Center*

Gerald Luecke, MSEE
Mgr. Technical Products Development

Charles W. Battle, Editor

Staff of Texas Instruments

Andrew Chisholm
Applications Engr., Bedford, England

Sebastino D'Arrigo
Engr. Mgr., Systems Dev., Rieti, Italia

Angelo Di Giovanni
Systems Engr., R&D Dept., Nice, France

Richard M. Pearson
Technical Writer, Bedford, England

Peter A. Roe
Telecomm. Specialist, Bedford, England

TEXAS INSTRUMENTS

P.O. BOX 225012, MS-54 ● DALLAS, TEXAS 75265

This book was developed by:

The Staff of the Texas Instruments Learning Center
P.O. Box 225012, MS-54
Dallas, Texas 75265

For marketing and distribution inquire to:

James B. Allen
Marketing Manager
P.O. Box 225012, MS-54
Dallas, Texas 75265

Acknowledgements:

The authors wish to acknowledge the tutelage and help of Dr. John Bellamy, whose knowledge of this subject fills his students and this book. They are grateful also for the friendship and advice of John McNamara of Digital Equipment Corporation, Ken Bean of Texas Instruments, and of Leo Goeller, who always has time to chat. Finally, we are grateful to the members of the International Communications Association, whose encouragement and support have changed our lives.

Figures reprinted with permission of Bell Telephone Laboratories and AT&T came from the list that follows. Other publications are credited individually.
1. Members of the Technical Staff, Bell Telephone Laboratories, *Engineering and Operations of the Bell System*, Bell Telephone Laboratories, Inc., 1977
2. Members of the Technical Staff, Bell Telephone Laboratories, *A History of Engineering and Science in the Bell System, The Early Years (1875-1925)*, Bell Telephone Laboratories, Inc., 1975
3. Members of the Technical Staff, Bell Telephone Laboratories, *Bell Laboratories Record*, Nov. 1980
4. Network Planning Division of AT&T, *Notes on the Network*, American Telephone and Telegraph Co., No. 500-029, 1980

Word Processing: *Design and artwork by:*

Betty Brown   Plunk Design

ISBN 0-89512-056-9
Library of Congress Catalog Number: 82-074287

Table of Contents

Preface

The public switched telephone network in the United States of America is one of the true wonders of the modern world. It provides the ability to interconnect any two out of more than one hundred million telephones, usually within a few seconds of the request for connection. It is controlled by the world's largest network of interconnected and cooperating computers. Yet the telephones in this network are usable by unskilled operators without formal training (almost any child of four or five can make a telephone call).

This book is in part about that network and the technology that has made it possible. But it is more about revolutionary changes that are taking place in the way telephone conversations and data are taken into the network, switched, and transmitted. Some of the technology still in everyday use in the telephone network dates from the decade of the invention of the telephone in the 1870's. However, even this old and traditional business is being forced by economics, regulation, and competition to make massive changes in the way it does business and in the equipment and techniques used to provide the telephone service.

We begin by presenting the fundamentals of the telephone network: how it began, what the components are, and how they are connected together. Next, we review the basic non-electronic telephone set. We then consider the effect of microelectronics on the construction and operation of the telephone set; for example, the effect on functions such as speech signal processing and interface with the telephone line, pulse and tone generation for dialing, and ringers. Next we discuss how microcomputers use digital techniques and stored programs to enhance the performance and features of the telephone set. Digital transmission techniques, electronics in the central office, and network transmission concepts and fundamentals are explained. Finally, we briefly consider the relatively new cordless telephone set and a new concept for mobile radio telephone service.

Like other books in the series, this book builds understanding step-by-step. Try to master each chapter before going on to the next one. A quiz is provided at the end of each chapter for self-evaluation of what has been learned. Answers are given at the back of the book.

The business of providing the equipment and service for both local and long distance telephone communication is today undergoing some of the most fundamental changes ever required to be made by any U.S. institution. The largest company in that business is, until January 1, 1984, the largest company on earth, and its product is a service that has become a necessity for all of us. Understanding the technical side of the telephone is necessary in understanding a force in modern life which has been and will continue to be as much an agent for change as the automobile, the airplane, and the computer.

J.F.
G.F.

The Telephone System

INTRODUCTION

The telephone arrived as a practical instrument over a century ago in 1876, an outgrowth of experiments on a device to send multiple telegraph signals over a single wire. Alexander Graham Bell, a native of Scotland, while conducting electrical experiments spilled acid on his trousers. His sulphurous reaction, the now famous "Mr. Watson, come here, I want you", brought Thomas A. Watson on the run not only because of his employer's distress, but because the words had been carried by electricity into Watson's room and reproduced clearly on his receiving set. The simple instrument being tested on Court Street in Boston on March 10, 1876 wasn't very practical (the acid was used in the system), but improvement followed so rapidly that putting into action Bell's concept of a public telephone network—"this grand system", "whereby a man in one part of the country may communicate by word of mouth with another in a distant place"—was well underway by January of 1878, when the first commercial exchange was operated in New Haven. By 1907, one hotel alone (the Waldorf Astoria in New York City) had 1,120 telephones and processed 500,000 calls per year.

That concept has grown into an industry in which one entity (American Telephone and Telegraph Company) is the largest company on earth, providing over one hundred million telephone sets, making a profit of several billion dollars per year, and employing over one million people.

This book will attempt to explain the devices that make the network what it is, how they work, why they are made the way they are, and how they are changing with the advent of digital circuits and microelectronics.

THE TELEPHONE SET

Telephone sets like those used to originate and receive telephone calls are shown in *Figure 1-1*. It is simple in appearance and operation yet it performs a surprising number of functions. Here is a list of the most important ones:

1. It requests the use of the telephone system when the handset is lifted.
2. It indicates that the system is ready for use by receiving a tone, called the dial tone.
3. It sends the number of the telephone to be called to the system. The number is initiated by the caller by pressing number keys or rotating a dial.
4. It indicates the state of a call in progress by receiving tones indicating the status (ringing, busy, etc.)

5. It indicates an incoming call to the called telephone by ringing bells or other audible tones.
6. It changes speech of a calling party to electrical signals for transmission to a distant party through the system. It changes electrical signals received from a distant party to speech for the called party.
7. It automatically adjusts for changes in the power supplied to it.
8. It signals the system that a call is finished when a caller "hangs-up" the handset.

Of course, for a telephone to be of any use, it must be connected to another telephone. In the very early days of telephony, the phones were simply wired together with no switching. This became impractical as the number of phones increased and the local exchange or central office was established to handle the switching and other functions.

a. Rotary Dial

b. Keypad Dial

Figure 1-1. *Telephone Set*
(Courtesy of Radio Shack)

THE LOCAL LOOP

Each subscriber telephone is connected to a central office that contains switching equipment, signaling equipment and batteries that supply direct current to operate the telephone as shown in *Figure 1-2*. Each phone is connected to the central office through a local loop of two wires called a wire pair. One of the wires is called T (for tip) and the other is called R (for ring) which refers to the tip and ring parts of the plug used in manual switchboards.

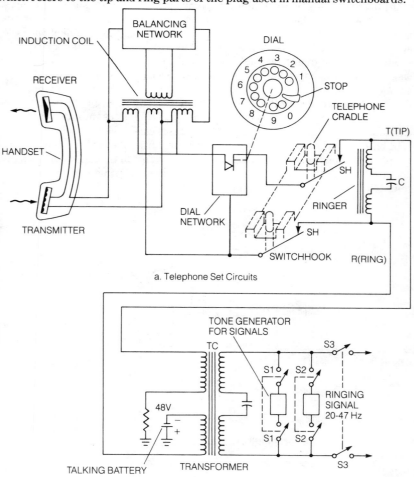

a. Telephone Set Circuits

b. Central Office Circuits

Figure 1-2. *Telephone Set and Central Office Exchange Simplified Circuits (Source: D.L. Cannon and G. Luecke,* Understanding Communications Systems, *Texas Instruments Incorporated, 1980)*

Switches in the central office respond to the dial pulses or tones from the telephone to connect the calling phone to the called phone. When the connection is established, the two telephones communicate over transformer coupled loops using the current supplied by the central office batteries.

Initiating a Call

When the handset of the telephone is resting in its cradle, the weight of the handset holds the switchhook buttons down and the switches are open. This is called the on-hook condition. The circuit between the telephone handset and the central office is open; however, the ringer circuit in the telephone is always connected to the central office as shown in *Figure 1-2*. The capacitor, C, blocks the flow of dc from the battery, but passes the ac ringing signal. (The ringer circuit presents a high impedance to speech signals so it has no effect on them.)

When the handset is removed from its cradle, the spring-loaded buttons come up and the switchhook closes. This completes the circuit to the exchange and current flows in the circuit. This is called the off-hook condition. (The on-hook, off-hook, and hang-up terms came from the early days of telephony, when the receiver was separate and hung on the switchhook when not in use as shown in *Figure 1-3*. This also explains why many people still refer to the handset of today as the receiver.)

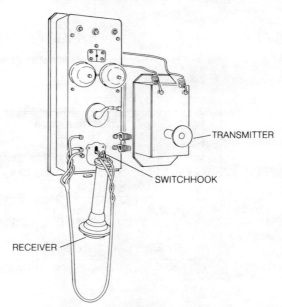

Figure 1-3. Early Telephone with Separate Receiver Hanging on Switchhook

The off-hook signal tells the exchange that someone wants to make a call. The exchange returns a dial tone to the called phone to let the caller know that the exchange is ready to accept a telephone number. (The telephone number also may be referred to as an address.)

Sending a Number

Some telephone sets send the telephone number by dial pulses while others send it by audio tones.

Dial Pulsing

Most telephone sets that use dial pulsing have a rotary dial as shown in *Figure 1-2*, which opens and closes the local loop circuit at a timed rate. The number of dial pulses resulting from one operation of the dial is determined by how far the dial is rotated before releasing it.

Dual Tone Multifrequency (DTMF)

Some telephone sets use the newer method of using audio tones to send the telephone number. These can be used only if the central office is equipped to process the tones. Instead of a rotary dial, these telephone sets have a push-button keypad with 12 keys for the numbers 0 through 9 and the symbols * (asterisk) and # (octothorpe). Pressing one of the keys causes an electronic circuit in the keypad to generate two output tones that represent the number.

Connecting the Phones

The central office has various switches and relays that automatically connect the calling and called phones. For now, assume that the connection has been made. The actual operation of switching systems will be covered in more detail a little later.

If the called phone handset is off-hook when the connection is attempted, a busy tone generated by the central office is returned to the calling phone. Otherwise, a ringing signal is sent to the called phone to alert the called party that a call is waiting. At the same time, a ringback tone is returned to the calling phone to indicate that the called phone is ringing.

Ringing the Called Phone

Early telephone circuits were point-to-point (not switched), and the caller gained the attention of the party at the other end by picking up the transmitter and shouting "Hello" or "Ahoy". This was not very satisfactory, and schemes based on a mechanical signaling arrangements were soon invented. The one in common use today, called the "polarized ringer", or bell, was patented in 1878 by Thomas A. Watson (Mr. Bell's assistant).

Answering the Call

When the called party removes the handset in response to a ring, the loop to that phone is completed by its closed switchhook and loop current flows through the called telephone. The central office then removes the ringing signal and the ringback tone from the circuit.

Talking

The part of the telephone into which a person talks is called the transmitter. It converts speech (acoustical energy) into variations in an electric current (electrical energy) by varying or modulating the loop current in accordance with the speech of the talker.

The part of the telephone that converts the electric current variations into sound that a person can hear is called the receiver. The signal produced by the transmitter is carried by the loop current variations to the receiver of the called party. Also, a small amount of the transmitter signal is fed back into the talker's receiver. This is called the sidetone.

Sidetone is necessary so that the person can hear his/her own voice from the receiver to determine how loudly to speak. The sidetone must be at the proper level because too much sidetone will cause the person to speak too softly for good reception by the called party. Conversely, too little sidetone will cause the person to speak so loudly that it may sound like a yell at the receiving end.

Ending the Call

The call is ended when either party hangs up the handset. The on-hook signal tells the central office to release the line connections. In some central offices, the connection is released when either party goes on-hook. In others, the connection is released only when the calling party goes on-hook.

Beyond the Local Loop

Thus far, the discussion of connecting two telephones together has been limited to local loops and a central office exchange. Most central office exchanges can handle up to 10,000 telephones. But what if it is required to connect more phones than 10,000, or connect phones in different cities, or in different states, or in different countries. Over the years, a complex network of many telephone exchanges has been established to accomplish these requirements. Let's look next at how this network is arranged.

THE PUBLIC SWITCHED TELEPHONE NETWORK

Exchange Designations

Each telephone exchange in North America has two designations, office class and name, to identify it and to describe its function. These are shown in *Figure 1-4*.

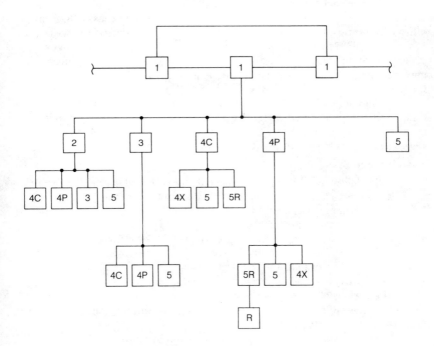

Key	Class	Name	Abbreviation
1	1	Regional Center*	RC
2	2	Sectional Center*	SC
3	3	Primary Center*	PC
4	4C	Toll Center	TC
4P	4P	Toll Point	TP
4X	4X	Intermediate Point	IP
5	5	End Office	EO
5R		End Office with Remote Switching Unit	
R		Remote Switching Unit	RSU

*May be a "point" rather than a "center". The abbreviation is then RP, SP, or PP.

***Figure 1-4.** Network Heirarchy*

Subscriber telephones are normally, but not exclusively, connected to End Offices. Toll (long distance) switching is performed by Class 4, 3, 2, and 1 offices. The Intermediate Point, or Class 4X office, is a relatively new class. It applies to all-digital exchanges to which remote unattended exchanges (called Remote Switching Units) can be attached. These Class 4X offices may interconnect subscriber telephones as well as other Class 5 and Class 4 exchanges.

The ten Regional Centers (Class 1 offices) in the U.S. and two in Canada all are connected directly to each other with large-capacity trunk groups. In 1981, there were 67 Class 2; 230 Class 3; 1,300 Class 4 and about 19,000 Class 5 exchanges in North America.

Interconnection

The network is organized like a tree, or rather like a small grove of trees, whose roots have grown together. *Figure 1-4* shows this in simplified form. Each exchange is optimized for a particular function. A call requiring service which cannot be performed by a lower class exchange is usually forwarded to the next higher exchange in the network for further processing.

The Regional Center, like the base of each tree, forms the foundation of the network. The branch levels are the Class 2, 3, 4, 4X, and Class 5 offices. Most offices are connected to more than one other, and the interconnections among the various offices are not as simple as shown in *Figure 1-4*. The interconnections depend on the patterns of the traffic arriving at and leaving each office.

The network makes connections by attempting to find the shortest path from the Class 5 office serving the caller to the Class 5 office serving the called party. The high-usage interoffice trunk groups which provide direct connection between offices of equal or lower level are used first. If they are busy, trunk groups at the next higher level (called final groups) are used. Digital logic circuits in the common control of each exchange make decisions based on rules stored in memory that specify which trunk groups are to be tried and in what order. These rules, for example, prevent more than nine connections in tandem, and prevent endless loop connections (called ring-around-the-rosy).

Structure

The supervisory signals used to set up telephone connections and the voice signals of the conversations are carried by transmission systems over paths called facilities. These systems are divided into three broad categories: Local, Exchange Area, and Long-Haul as shown *in Figure 1-5, 1-6 and 1-7*.

---

[1]Nine tandem connections have never been known to occur.

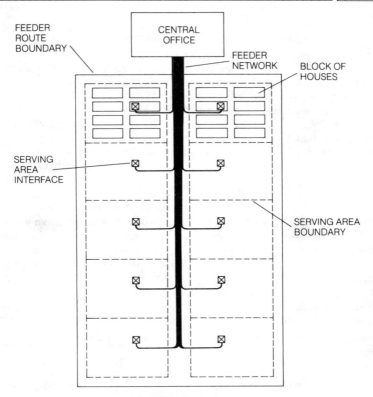

a. Local Distribution Area

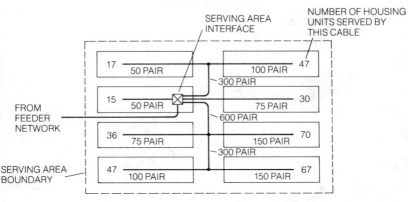

b. Detail of a Serving Area

Figure 1-5. *Local Network*
(Courtesy of Bell Laboratories)

The Local Network

The local network shown in *Figure 1-5* is the means by which telephones in residences and businesses are connected to central offices. The local facilities are almost exclusively wire pairs which fan out like branches of a tree from a point called the wire center throughout a serving area. Serving areas vary greatly in size, from an average of 12 square miles in urban locations to 130 square miles for rural areas. More than one central office is often required for a serving area in urban areas, but one central office is usually sufficient in rural areas. An average wire center in an urban area will serve 41,000 subscriber lines and 5,000 trunks. The urban exchanges are generally of higher call carrying capacity than the rural exchanges.

The Exchange Area Network

The exchange area network is intermediate between the local network and the long-haul network. A simplified example is shown in *Figure 1-6*. Exchanges are interconnected with exchange area transmission systems. These systems may consist of open wire pairs on poles, wire pairs in cables, microwave radio links, and fiber optic (light pipe) cables. The exchange area network normally interconnects local exchanges and tandem exchanges. Tandem exchanges are those that make connections between central offices when an interoffice trunk is not available. A tandem exchange is to central offices as a central office is to subscriber telephone sets.

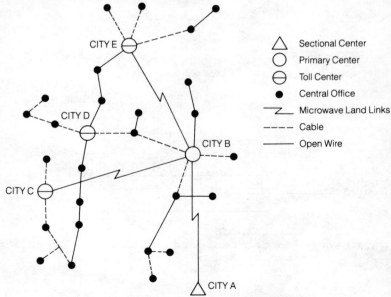

Figure 1-6. *Exchange Area Network*
(Courtesy of Bell Laboratories)

The Long-Haul Network

In the long-haul network, local exchanges are interconnected with toll (long distance) exchanges. A simplified example is shown in *Figure 1-7*. These facilities are normally of high capacity per circuit, and consist mostly of cable and microwave radio links. In some paths (called routes) which require a great many channels, such as the backbone routes between Boston and Washington, very high capacity fiber optic cables are being installed. This new Northeast corridor system consists of several fiber optic links, each carrying about 4,000 voice channels simultaneously. Such high capacity links are expected to save operating telephone companies $50 to $100 million in construction and operating costs by 1990.

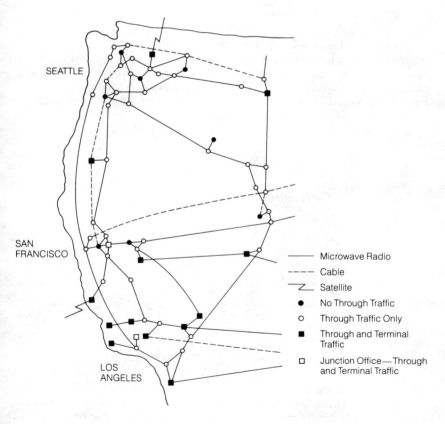

Figure 1-7. *Long Haul Network*
(Courtesy of Bell Laboratories)

TYPES OF TRANSMISSIONS

Spoken messages or voice signals are not the only signals that are transmitted down a telephone line. In the previous discussion of making a connection between the calling telephone and the called telephone, some of these other signals were discussed: dial tone, dial pulses or key tones used for sending a number, busy tone and ringback tone. These are for control of the switching connections or to indicate the status of the call. Such signals are called control signals or supervisory signals. They may be tone signals (analog) or ON-OFF (digital) signals. Therefore, if one were to examine the signals on many local loops, one would find analog voice signals, analog tone signaling, and digital ON-OFF signaling. It would be a mixture of analog and digital signals.

Analog Voice Transmissions

Signals that have continuously and smoothly varying amplitude or frequency are called analog signals. Speech (or voice) signals are of this type. They vary in amplitude and frequency. *Figure 1-8* shows the typical distribution of energy in voice signals. The vertical axis is relative energy and the horizontal axis is frequency. It shows that the voice frequencies that contribute to speech can extend from below 100 hertz to above 6,000 hertz. However, it has been found that the major energy necessary for intelligible speech is contained in a band of frequencies between 200 hertz and 4,000 hertz.

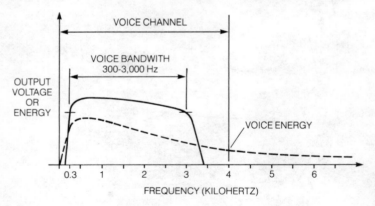

Figure 1-8. Voice Energy Versus Frequency

Voice Channel Bandwidth

In order to eliminate unwanted signals (noise) that could disturb conversations or cause errors in control signals, the circuits that carry the telephone signals are designed to pass only certain frequencies. The range of frequencies that are passed are said to be in the pass band. Zero to 4,000 hertz is the pass band of a telephone system voice channel—a VF Channel.

(Sometimes this band is called a message channel.) Bandwidth is the difference between the upper limit and lower limit of the pass band; thus, the bandwidth of the VF channel is 4,000 Hz. However, not all of the VF channel is used for the transmission of speech. The voice pass band is restricted to 300 to 3,000 hertz as shown in *Figure 1-8*. Hence, any signal carried on the telephone circuit which is within the range of 300 to 3,000 hertz is called an in-band signal as shown in *Figure 1-9*. Any signal which is not within the 300 to 3,000 Hz band, but is within the VF channel, is called an out-of-band signal. All speech signals are in-band signals. Some signaling transmissions are in-band and some are out-of-band.

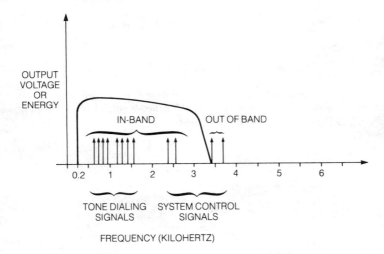

Figure 1-9. *In-Band and Out-of Band Signaling*

Voice Channel Level

The loudness or amplitude of signals on telephone circuits is usually referred to as the level of the signal. The level of a signal is expressed in terms of the power which the signal delivers to a load. For example, a pair of telephone wires running together in a cable form a transmission line with an impedance of 600 ohms. Impedance is to ac circuits what resistance is to dc circuits. As shown in *Figure 1-10*, the power delivered to a balanced pair transmission line is:

$$P_{load} = \frac{e_s^2}{Z}$$

Where:
$$P_{load} = \text{Power in watts}$$
$$e_s = \text{Signal level in volts}$$
$$Z = \text{Impedance in ohms}$$

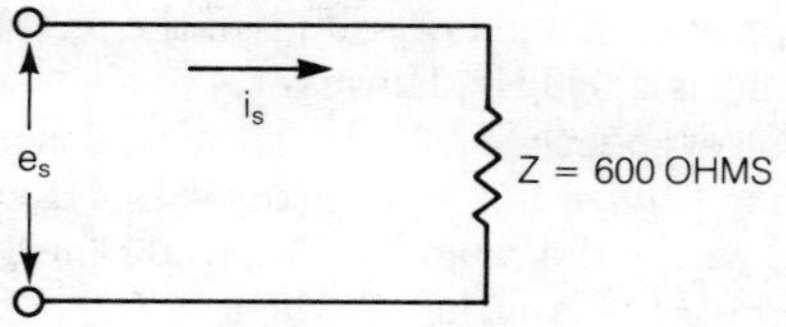

$$e_s \quad i_s \quad Z = 600 \text{ OHMS}$$

$$P_{load} = e_s \times i_s \qquad i_s = \frac{e_s}{Z} \qquad \begin{aligned} P_{load} &= \text{Power in watts to load} \\ e_s &= \text{Signal level in volts} \\ i_s &= \text{Signal current in amperes} \\ Z &= \text{Impedance in ohms} \end{aligned}$$

$$= e_s \times \frac{e_s}{Z}$$

$$= \frac{e_s^2}{Z}$$

Figure 1-10. *Power Delivered to Wire Pair*

Signal level usually is expressed relative to some reference. In telephone circuits and audio circuits, the reference level is 1 milliwatt of power to the load. If P_{load} = 1 milliwatt (0.001 watt) and Z = 600 ohms then, as shown in *Figure 1-10*;

$$1\text{mW} = \frac{e_s^2}{600 \text{ ohms}}$$

or

$$600 \times 1 \times 10^{-3} = e_s^2$$
$$0.6 = e_s^2$$
$$0.775 = e_s$$

Therefore, a signal level of 0.775 volt applied across 600 ohms produces 1 milliwatt of power.

Analog signals that are transmitted at a constant frequency also can have their level expressed in decibels (dB). It is another means of expressing the signal power delivered to a load. In technical terms as an equation:

$$dB = 10 \log_{10} (P_1/P_2)$$

It is a shorthand way of expressing the ratio of power P_1 to power P_2. *Table 1-1* lists some of these ratios.

A special decibel ratio is established when 1 milliwatt is used as the reference power, P_2. Under this condition, the dB power ratio is classified as measured in dBm (decibels referenced to 1 milliwatt). Therefore, from *Table 1-1*, if P_2 = 1 milliwatt, then a signal at 0 dBm will be delivering a power, P_1, of 1 milliwatt to the load because the ratio of P_1 to P_2 must be 1. Said another way, when a signal produces a power P_1 into a load of 600 ohms that has a 20 dBm level, it is delivering 100 milliwatts of power (P_1) compared to the reference power of 1 milliwatt for P_2.

Table 1-1. *Power Ratios in dB*

dB	P1/P2
40	10,000
30	1,000
20	100
10	10
3	2.0
0	1
−3	0.5
−10	0.1
−20	0.01
−30	0.001
−40	0.0001

In telephone systems, the 0 dBm level is usually set at the sending end of a transmission line at the output of the switch. This point then becomes a system reference point called the zero transmission level point (0 TLP). Once the 0 TLP is chosen and the 0 dBm level applied at that point, all other power gains and losses in the transmission path between that point and the next switch output can be measured directly with respect to the 0 TLP. If the signal magnitude is measured, then the unit dBm0 is used. If only the relative gain or loss is indicated, the unit dB is used.

Voice Channel Noise

Transmission systems often must operate in the presence of various unwanted signals (referred to generally as noise) that distort the information being sent. Lightning, thermal noise, induced signals from nearby power lines, battery noise, corroded connections, and maintenance activities all contribute to degradation of the signal. Analog channel speech quality is primarily determined by the absolute noise level on the channel when it is idle; that is, when no speech signal is present. Speech tends to mask any noise present, but noise in an idle channel is quite objectionable to a listener. Stringent standards (−69 dBm0 up to 180 miles and −50 dBm0 up to 3,000 miles with −16 dBm0 as speech level) have been set for this idle channel noise in the US network.

Another type of noise that originates from the voice transmission itself is an echo. The primary echo is the reflection of the transmitted signal back to the receiver of the person talking. The amount of delay in the echo depends on the distance from the transmitter to the point of reflection. The effect of the delay on the talker may be barely noticeable to very irritating to down right confusing. Echo also affects the listener on the far end, but to a lesser degree. Echos are caused by mismatches in transmission line impedances which usually occur at the hybrid interface between a 2-wire circuit and a 4-wire transmission system. The effect of echo is reduced by inserting loss in the lines.

Multiplexing

A local loop can carry only one voice channel conversation at a time. This is not economical for toll transmission and a method was devised so that a transmission path can carry many telephone conversations at the same time. This is accomplished by multiplexing. For analog signals, frequency division multiplexing (FDM) is used. In simplified terms, this means that several telephone conversations are all sent together over one transmission channel, but are separated by their frequency.

The basic principles of this are shown in *Figure 1-11* and *1-12*. In *Figure 1-11*, a voice signal having frequencies within the voice frequency channel bandwidth of from 0 to 4 kilohertz is changing the amplitude of another frequency (8,140 kilohertz in this case) which is called the carrier frequency. The 0-4 kHz voice frequency signal is amplitude modulating the 8,140 kHz carrier. The information in the voice signal is being carried by the changing amplitude of the 8,140 kHz signal and the voice frequencies have been translated to different frequencies.

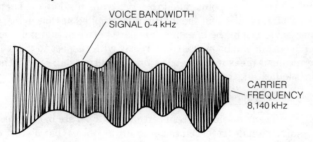

VOICE BANDWIDTH
SIGNAL 0-4 kHz

CARRIER
FREQUENCY
8,140 kHz

Figure 1-11. *Carrier Modulated by Voice Signal*

If different voice signals (different telephone conversations) are placed on different carrier frequencies, then many conversations may be multiplexed on one transmission path and transmitted to the receiving point. At the receiving point, the different conversations can be identified and separated by their unique frequency, and the original conversation can be recovered from the carrier (demodulated) and sent to the called telephone.

The multiplexing of the signals is shown in *Figure 1-12* for 12 voice channels. Since each voice channel has a 4 kHz bandwidth, 12 channels require 4 x 12 or 48 kHz bandwidth. Since the lower frequency in this example is 8, 140 kHz, the output multiplexed signal frequency extends from 8,140 kHz to 8,188 kHz (8,140 + 48 = 8,188). It should be apparent that if the individual voice channel bandpass were made larger, the spread in carrier frequency would have to be larger; or if the number of voice channels to be multiplexed together were increased, the spread in carrier frequencies would need to be larger. In technical terms, in general, as the number of voice channels to be transmitted over a transmission path increases, the required bandwidth of the transmission path must increase.

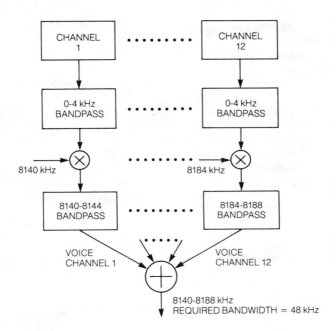

Figure 1-12. *Multiplexed Frequency Signals*
(Source: D.L. Cannon and G. Luecke, Understanding Communications Systems,
Texas Instruments Incorporated, 1980)

Signaling Transmission

As stated previously, signaling refers to specific signals on the
transmission line that are used for controlling the connection from the calling
telephone to the called telephone, or that are used to indicate the status of a
call as it is being interconnected. The first type to be discussed is dc signaling.

DC Signaling

DC signaling is based on the presence or absence of circuit current or
voltage, or the presence of a given voltage polarity. The state of the signal
indicates on-hook, off-hook, dial pulses, or status of the interconnection. These
signals are ON-OFF type digital signals.

On local loops, on-hook is indicated by an open circuit and no current
flow. Off-hook is signaled by a closed circuit and continuous current flow. Dial
pulses consist of current flow interrupted at a specified rate as discussed
previously. (A potential problem with dc signaling is that dial pulses spaced
too far apart may be mistaken for an on-hook signal by the exchange.
However, due to careful design, this problem does not occur very often.)

A type of dc signaling called reverse battery signaling is used between central offices to indicate the status of the switched connection. When the near end exchange requests service, an idle trunk is seized. A polarity of a given voltage exists on the trunk which indicates to the near end that the called phone is on hook and ringing. The far end exchange acknowledges and indicates to the near end that the called party has answered by reversing the voltage polarity.

E&M signaling is used for the same purpose on long interoffice and short-haul toll trunks. This type signaling requires two extra wires in the originating and terminating trunk circuits, one for the E lead and the other for the M lead. Since separate wires are used for each, the on-hook and off-hook states can be signaled from both ends of the circuit as shown in *Table 1-2*. This allows signaling to be sent in both directions at the same time without interfering with one another. Sometimes two wires are used for each signal to avoid noise problems caused by a common ground.

Table 1-2. *E&M Signaling*

State	E-Lead (Inbound)	M-Lead (Outbound)
On-hook	Open	Ground
Off-hook	Ground	Battery Voltage

Tone Signaling

Various tones are used for both control and status indication. The tones may be single frequency or combinations of frequencies. These are analog signals that are either continuous tones or tone bursts (tones turned on and off at various rates). The call progress tones listed in *Table 1-3* are sent by the exchange to the calling phone to inform the caller about the status of the call. For example, the dial tone, which has been mentioned previously, is a continuous tone made by combining the frequencies of 350 Hz and 440 Hz. The busy signal that tells the caller that the called telephone is busy (off-hook) is a combination frequency tone that appears in bursts of 0.5 second on time separated by an off time of 0.5 second. The receiver off-hook warning signal is a combination frequency tone of four frequencies which is on for 0.1 second and off for 0.1 second. This signal is very loud in order to get the attention of someone to "hang up" the receiver (handset) that has been left off-hook. All of these tones, as well as the DTMF addressing tones discussed previously, are in-band signaling.

Tone signaling between exchanges may be in-band or out-of-band. The most commonly used single frequency (SF) tones are 2,600 Hz for in-band and 3,700 Hz for out-of-band signaling. E&M signals are converted to an SF tone for transmission on carrier systems because the dc signals cannot be

Table 1-3. *Network Call Progress Tones*

Tone	Frequency (Hz)	On Time (Sec.)	Off Time (Sec.)
Dial	350 + 440	Continuous	
Busy	480 + 620	0.5	0.5
Ringback, Normal	440 + 480	2	4
Ringback, PBX	440 + 480	1	3
Congestion (Toll)	480 + 620	0.2	0.3
Reorder (Local)	480 + 620	0.3	0.2
Receiver Off-hook*	1400 + 2060 + 2450 + 2600	0.1	0.1
No Such Number	200 to 400	Continuous, Frequency modulated at 1 Hz Rate	

*Receiver off-hook is a very loud tone, 0 dBm per frequency.

transmitted. The tone indicates on-hook when present and off-hook when not present. Multifrequency (MF) supervisory signaling uses six frequencies: 700; 900; 1,100; 1,300; 1,500 and 1,700 Hz for transmitting address information (the telephone number) over toll facilities. The frequencies are used in pairs to represent the numerals 0 through 9 and some control functions much like DTMF is used at the telephone set.

Control Signals that are Digital

Instead of just interrupting a dc voltage, as in the case of dc signaling, or interrupting continuous tones to provide tone bursts, control or supervisory signals also can be digital codes. Instead of being ON-OFF signals that occur at random times, they are combinations of signals that have two levels, 0 and 1, and that have a definite time relationship with each other. This is illustrated in *Figure 1-13*. In the telephone system, the binary digit (bit) 1 and 0 levels shown may be represented by voltage or current levels. Note that the bits occur in a particular time sequence. For example, in *Figure 1-13*, a binary code of 8 bits is shown with bits d_0 through d_7 always occurring in the same time slot, t_1 through t_8, when transmitted in sequence. For a particular system design, once the time relationships of when the bits occur is set, the time relationship doesn't change.

The control information can be contained in the binary code in several ways. All 8 bits may be used as a group to represent a number from 0 to 255. The binary code for the number 234 is shown across the first line in *Figure 1-13*. On the left side, the code is presented in 1's and 0's, and on the right side, the code is presented as voltage levels or pulses. Or the 8-bit group may represent a letter of the alphabet in a data communications code. A letter T in ASCII is shown on the second line. Or individual bits or subgroups of the 8-bit code may be used to command different functions. Examples of subgroup codes for the functions A, B and C are shown on the third line.

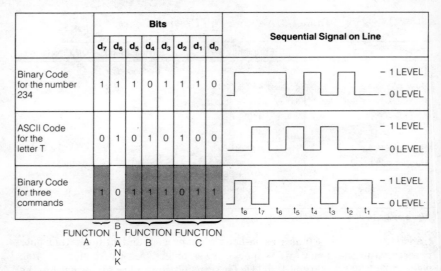

	Bits								Sequential Signal on Line
	d_7	d_6	d_5	d_4	d_3	d_2	d_1	d_0	
Binary Code for the number 234	1	1	1	0	1	1	1	0	
ASCII Code for the letter T	0	1	0	1	0	1	0	0	
Binary Code for three commands	1	0	1	1	1	0	1	1	t_8 t_7 t_6 t_5 t_4 t_3 t_2 t_1

FUNCTION A BLANK FUNCTION B FUNCTION C

Figure 1-13. Serial Digital Control Signals

Common Channel Interoffice Signaling

All the signaling methods discussed so far send the control and addressing signals over the same circuit as the voice signals. Another method that is used separates the control signals from the voice signals. The control signals are sent over a separate circuit where they are detected and do the control and switching of lines independently from the voice signals. This is called Common Channel Interoffice Signaling (CCIS). CCIS is illustrated in *Figure 1-14*. The basic control is by digital computer and CCIS is a separate data network for exchanging control signals among these computers. As the

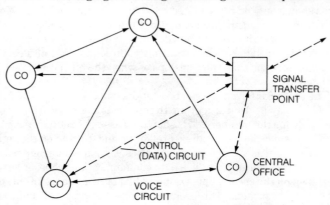

Figure 1-14. Common Channel Interoffice Signaling (CCIS)

name suggests, this method of signaling is used on the interconnecting trunks that carry signals between central offices.

Digital Transmission

Many advances have been made in solid-state electronics and integrated circuits that handle digital signals, so that high functional density integrated circuits with expanded signal handling capability can be put into a small space, are low cost, operate with low power and have long term reliability. Because of this, telephone system designs are changing toward an all digital network. The newer systems convert voice and signaling information to digital signals for transmission. When both voice and signaling information are transmitted in digital form it is called digital transmission.

Instead of the voice signal being processed as an analog signal, it is converted into a digital signal and handled with digital circuits throughout the transmission process. When it arrives at the central office that serves the called telephone, it is converted back to an analog signal to reproduce the original voice transmission. (In the future, the digital signal may travel all the way to the telephone set.)

When the binary signal is transmitted in serial form as shown in *Figure 1-13*, and the code varies as the signal changes, the method is called pulse code modulation (PCM). Because the voice signals and the supervisory signals are low-frequency signals (300-3,000 Hz), and digital circuits can operate at very high frequencies (millions of cycles per second), voice signals from many conversations can be sent in series on the same line. This is called time division multiplexing (TDM).

For digital transmission, multiplexing is done in a particular way as shown in *Figure 1-15*. To illustrate the technique, suppose a person is located at point A and can see the binary codes of *Figure 1-15*. The binary codes pass by point A serially one bit at a time. In this case, there are eight bits (d0 to d7) in each code. The value of the signal on Channel 1 is represented by the combination of 1's and 0's in the 8-bit code for Channel 1. The value of the signal on Channel 2 is represented by the combination of 1's and 0's in the 8-bit code for Channel 2. Channel 2's code follows Channel 1's code. Instead of the transmitted signal continuously representing the signal of one channel as on an analog circuit, the signals from all 24 channels are mixed together, but in a definite pattern. Channel 2 is multiplexed behind Channel 1, Channel 3 behind Channel 2, and so on until the codes for 24 channels have been multiplexed together, one following the other in serial fashion in time as shown. So the observer at point A would see 8-bits for Channel 1, 8-bits for Channel 2, 8-bits for Channel 3 and so on until Channel 24 came by. The pattern would then repeat beginning again with Channel 1. This is the way that many conversations that have been digitally encoded by PCM are digitally transmitted by TDM over one channel. TDM and PCM will be covered in much more detail in later chapters.

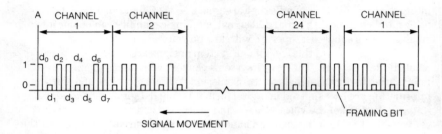

Figure 1-15. *Multiplexed PCM Signals*
(Source: D.L. Cannon and G. Luecke, Understanding Communications Systems, *Texas Instruments Incorporated, 1980)*

SWITCHING SYSTEMS

Now that there is some understanding of the transmission of the signals, let's look at how the called telephone and the calling telephone are actually connected by selecting one of a multitude of paths. Several mechanisms have been used to provide control of the process of switching one circuit among many others in the hundred-odd years since Mr. Bell envisioned "this grand system". The first, of course, was manual.

Manual Control

Early telephone switchboards were operated manually using a jack for each line and two plugs on a long flexible wire, called a cord pair *(Figure 1-16a)*, for making the connection. The cords pairs appeared in rows on a shelf in front of the operator, and the jacks (called line appearances) were mounted on a vertical panel as shown in *Figure 1-16b*. To make a connection, the operator picked up a cord *(Figure 1-16c)*, plugged it into the jack corresponding to the line requesting service, obtained from the calling party the name or number of the desired party, then plugged the other end of the cord pair into the correct outgoing line jack. There are many thousands of cord switchboards still in operation, a tribute to the versatility and ease of programming of the control system, and the personal touch it provides. However, not every subscriber appreciated the personal touch.

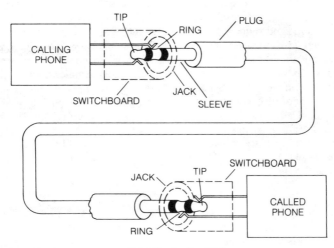

a. Patch Cord or Cord Pair

b. Board with Patch Cords
(Courtesy of Bell Laboratories)

c. Operator at Work
(Courtesy of Bell Laboratories)

Figure 1-16. *Manual Switchboards*

Progressive Control

In 1889, a Kansas City, Missouri undertaker, Almon B. Strowger, began to suspect that potential clients who called the operator of the local manual exchange and requested "an undertaker" were more often than not being connected to a firm down the street. This suspicion was reinforced when he learned that the telephone operator was the wife of the owner of the other funeral parlor in town. In the best tradition of pioneer America, Mr. Strowger invented a mechanical substitute for the biased operator which could complete a connection under direct control of the calling party. This simple device is variously called the Strowger, two-motion, or step-by-step switch. It was patented in 1891 and became the basis of a very large fraction of the installed telephone switching systems in the world. As of 1978, 53% of the Bell System exchanges in service (over 23,000,000 subscribers) used Strowger switching, even though the Bell System did not begin installing Strowger switching until about 1918.

Step-by-Step

The Strowger, or step-by-step, switch connects pairs of telephone wires by progressive step-by-step operation of several series switches (called the switch train) operating in tandem. Each operation is under direct control of the dial pulses produced by the calling telephone. *Figure 1-17* shows what happens in simplified schematic form. The telephone line shown is actually a pair of wires. When the calling telephone goes off hook, current flowing in the local loop operates a relay in the exchange, causing the first switch in the train (the linefinder) to search for the active line by stepping vertically until the vertical contact is connected to the off-hook line. The line finder then steps horizontally until it finds a first selector that is not in use on another call. This is the next switch in the train. When a free first selector is connected, a dial tone is returned to the calling party. The first selector switch waits for the first digit to be dialed, then steps vertically one step for each dial pulse received. When it has taken in one digit, it steps horizontally until a free second selector is found, and the process is repeated. Thus, the first switch in the train (the linefinder) takes in no digits, the second and third takes in one digit each, and the last switch in the train (called a connector) takes in the final two digits. A 10,000 line exchange requires four digits to be dialed (0000 through 9999) and requires four switches for each connected call (linefinder, first selector, second selector, and connector). A line drawing of a Strowger switch bank is shown in *Figure 1-18*.

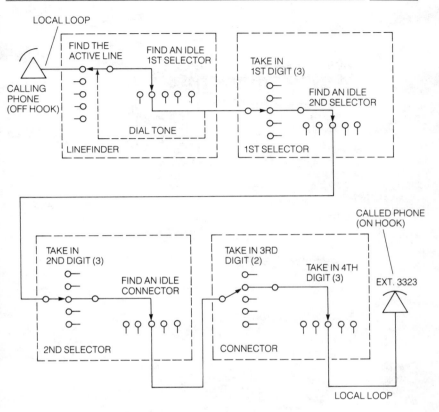

Figure 1-17. *Strowger Step-by-Step Switching*

The Strowger switching system has the following significant limitations:

1. Since several switches are operated in tandem and the switches (except for the first one) are shared among many incoming lines, it is possible for a call to become blocked partway through the dialing sequence, even though the called line is free.
2. It is not possible to use tone dialing (DTMF) telephones directly. (They may be used if the central office is equipped with a conversion device.)
3. The switch requires the successful sequential (step-by-step, time related) operation of several relays and a sizable voltage and current is switched each time a switch is stepped. Consequently, the mechanical reliability of the switches is low, they require large amounts of maintenance by skilled people, and they generate large amounts of electrical and mechanical noise.
4. Since the switching network is hard-wired, it is difficult to make changes in the switching arrangement.

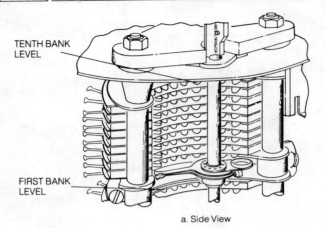

TENTH BANK
LEVEL

FIRST BANK
LEVEL

a. Side View

WIPER ASSEMBLY

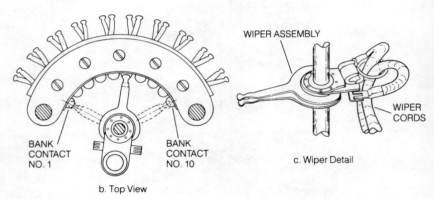

WIPER
CORDS

BANK
CONTACT
NO. 1

BANK
CONTACT
NO. 10

c. Wiper Detail

b. Top View

Figure 1-18. *Strowger Switch Bank*
(Courtesy of Bell Laboratories)

Common Control

Primarily due to the inflexibility and large maintenance costs
associated with Strowger-type switching networks, the concept of common
control was brought back, but with a new type of switching matrix called a
crossbar. The common control can be assigned to an incoming call as required.
It takes in the dialed digits, and then sets up the path through the switching
matrix according to hard-wired or stored-program rules. These rules provide
for variations in the handling of local and long distance calls, for choosing an
alternate route for a call in case the first route chosen is busy, and for trying
the call again automatically in case of blocking or faults in the switching path.
The common control element may be a relay operated device called a marker
or a stored-program controlled digital computer.

Crossbar

Crossbar, as the name implies, depends on the crossing or intersection of two points to make a connection. The switching matrix is shown in *Figure 1-19a*. It is called a crosspoint array. Its operation depends on energizing a vertical line and a horizontal line and the point where they intersect represents the connection made. Therefore, any one of the input lines shown (I1 through I6) can be connected to any one of the output lines (O1 through O10) by energizing a particular input line and a particular output line.

As shown in *Figure 1-19b*, the crossbar matrix is controlled by common control. Control signals from transmission lines are detected and used to control the matrix to connect the proper lines for the path from the calling telephone to the called telephone.

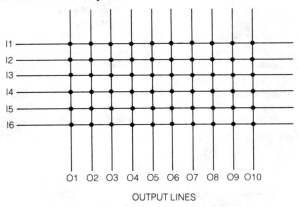

a. Crosspoint Array

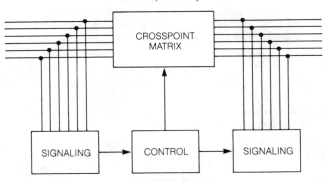

b. System Organization

Figure 1-19. *Crossbar Switching*
(From J. Bellamy, Digital Telephony, *John Wiley & Sons, Inc., 1982, Copyright© 1982 by John Wiley & Sons, Inc., Reprinted by permission of John Wiley & Sons, Inc.)*

Electromechanical Version

Figure 1-20 illustrates a relay type of mechanism that has vertical and horizontal selector bars operated by electromagnets which close relay contacts to provide the matrix interconnection. When a horizontal select magnet is energized, its horizontal selecting bar rotates slightly on its axis. This moves a selecting finger up or down to allow either an upper or lower bank of horizontal contacts to complete a circuit to the vertical contacts when the appropriate vertical select magnet is energized. The vertical select magnet moves the vertical holding bar sideways to push on the selecting finger to close the respective horizontal contacts to the vertical contacts. For the other horizontal select magnets that have not been energized, the associated selecting fingers are in the middle position and pass between the horizontal contacts when the vertical holding bar pushes on the selecting fingers. Therefore, no crossbar connection is made at any other point.

One crosspoint is at the intersection of each horizontal and vertical bar, as shown in *Figure 1-19a*. A crosspoint is provided for each wire of the wire pair so that both wires of the line are switched. Once connected, the switch path is maintained by the current flowing through the vertical select magnet coil. The horizontal select magnet is deenergized and the horizontal selecting bar returns to its idle position, but the previously selected finger is held by the vertical holding bar. When the calling end goes on hook (hangs up), the vertical select magnet current is interrupted and the crosspoints are released. Thus, the calling party controls the connection.

Reed Relays

Another type of switch uses reed relays to make the connections. The reed relay is a small, glass-encapsulated, electromechanical switching device as shown in *Figure 1-21*. These devices are actuated by a common control which selects the relays to be closed in response to the number dialed, and sends pulses through coils wound around the relay capsules. The pulses change the polarity of magnetization of plates of magnetic material fitted alongside the glass capsules. The contacts open or close in response to the direction of magnetization of the plates, which is controlled by the positive or negative direction of the pulse sent through the windings. Since the contacts latch, no holding current is required for this type of crosspoint, but separate action is required by the common control to release the connection (unlatch or reset the relay) when one party or the other hangs up.

Reed relays have improved the reliability and maintainability of switches a great deal. Crossbar switches still provide much of the switching for long distance or long haul telephone calls in the United States. In addition, reed relays are an important part of stored program controlled electronic switching systems.

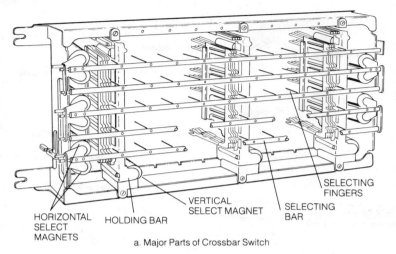

HORIZONTAL SELECT MAGNETS   HOLDING BAR   VERTICAL SELECT MAGNET   SELECTING BAR   SELECTING FINGERS

a. Major Parts of Crossbar Switch

Horizontal Select Magnets rotate selecting bar to position selecting finger either up or down. Horizontal Magnets release after vertical holding bar is positioned but selecting finger is held by vertical holding bar.

Selecting finger position to actuate upper horizontal contacts.

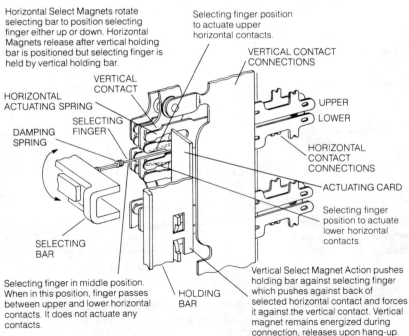

VERTICAL CONTACT CONNECTIONS

HORIZONTAL ACTUATING SPRING

VERTICAL CONTACT

SELECTING FINGER

DAMPING SPRING

UPPER
LOWER

HORIZONTAL CONTACT CONNECTIONS

ACTUATING CARD

Selecting finger position to actuate lower horizontal contacts.

SELECTING BAR

Selecting finger in middle position. When in this position, finger passes between upper and lower horizontal contacts. It does not actuate any contacts.

HOLDING BAR

Vertical Select Magnet Action pushes holding bar against selecting finger which pushes against back of selected horizontal contact and forces it against the vertical contact. Vertical magnet remains energized during connection, releases upon hang-up.

b. Detail of one Selecting Crosspoint

Figure 1-20. *Crossbar Switch*
(Reprinted with permission, abc TeleTraining, Inc., Geneva, Ill.)

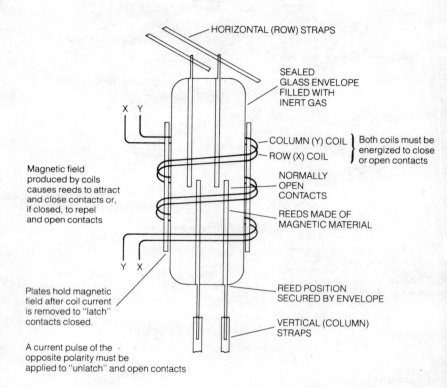

Figure 1-21. *Reed Relay Structure*

All of the step-by-step, crossbar, and reed relay switching is called space division switching because each telephone conversation is assigned a separate physical path through the telephone system. The PCM time division multiplexed digital transmissions discussed previously are different because they place many interleaved conversations onto one telephone line.

Local Loops and Trunks

As the telephone connection extends from the calling phone, it proceeds over the local loop to the first exchange, the central office. From the central office, it proceeds over trunk connections selected by the control signals to other exchanges. Control signals and functions different from those used in the local loop are required for making trunk interconnections. Let's examine several of these.

A dial tone is returned to the calling telephone when the local exchange is ready to receive the called number after recognizing that a telephone has been taken off-hook. To tell a trunk the same thing requires an appropriate trunk acknowledgement signal.

The acknowledgement signal may be a reversal of battery polarity, momentary reapplication of a tone, an interruption of the circuit (wink), or transmission of coded tone signals (proceed-to-send tone). The acknowledgement signal is returned only after the exchanges have made the proper connections to receive the called number.

The called number must be transmitted over the trunk just like the dial pulses or dial tones from the calling telephone. This is done with a sending device in the originating exchange. At the receiving exchange, the common control may be the crossbar control circuit, or on newer exchanges, it is an electronic circuit called an incoming register that receives and stores the digits of the called number. After all of the digits arrive, they are used by the common control section to direct the switching of the network to route the conversation over the correct path.

When an exchange has received enough information to complete a connection, it selects the appropriate outbound trunk or line, seizes it, and applies a signal to alert the distant end that a call has arrived and requires service. When the next exchange detects the call, it sends back an acknowledgement signal to the sending exchange to send the dialed number.

At the same time that the call connection is progressing toward the called phone, each exchange in the train sends back status signals to the previously connected exchange to indicate how the call is progressing.

When the call arrives at the local central office serving the called telephone, and if the called telephone is on-hook, the ringing voltage is sent over the local loop to ring the called telephone. At the same time, the central office serving the called telephone returns a ringback signal all the way back to the calling phone to indicate that the called telephone is ringing.

When the called telephone answers, its central office provides the power for the called phone and the conversation proceeds.

TRANSMISSION SYSTEM FACILITIES

The physical lines, the relays, switches, cables, power supplies, electronic circuits, transformers, impedance matching networks etc., are called the transmission system facilities. Whether voice only, or voice and analog signaling, or digital signaling, or all digital signals are sent over a particular path depends on the facilities. Also, the facilities determine the number of channels that can be sent over a particular path. Facilities generally fall into three categories: Metallic, Analog Carrier and Digital Carrier.

Metallic

Metallic facilities are wire pair line circuits carrying only one VF channel, and on which in-band or out-of-band signaling, including DC signaling, can occur. Typically, a metallic facility is used to connect a central office to the telephones in homes and small businesses.

Analog Carrier

Analog carrier facilities may operate over different media, such as wire lines, multi-wire cable, coaxial cable, microwave radio land or satellite links, or fiber optic cable. Each circuit carries from a few to several thousand individual analog VF channels and in-band analog or CCIS signaling is used.

Table 1-4 shows the common types of analog carrier facilities, their bandwidth and the number of voice channels that can be carried by each. Note that an L5 transmission path can carry 108,000 different conversations at the same time.

Table 1-4. *Analog Carrier and Multiplex Systems*

Medium	Carrier Type	Number of VF Channels	Number of Pairs or Radio Channels	Bandwidth
Open-Wire	O	4-12	2	200 kHz
	On-2	24	2	200 kHz
Twisted Pairs	K	12	2	300 kHz
In Cables	N-1	12	2	300 kHz
	N-2	12	2	300 kHz
	N-3	24	2	300 kHz
	N-4	24	2	300 kHz
Coaxial	L1	1,800	3	3 MHz
Cable Pairs	L3	9,300	5	10 MHz
	L4	32,400	9	20 MHz
	L5	108,000	10	68 MHz
Microwave	TD-2	19,800	11	500 MHz
Radio	TD-3	12,000	10	500 MHz
	TH-1	10,800	6	500 MHz
	TH-3	14,400	6	500 MHz
	TM-1	3,600	4	500 MHz
	TJ	1,800	3	1000 MHz
	TL-1	720	3	1000 MHz
	TL-2	2,700	3	1000 MHz
	AR6-A	42,000	7	500 MHz

One interesting application of analog multiplexing occurs in undersea telephone cables. The latest undersea cable, called the SG system, carries 4,000 channels in both directions on a single coaxial conductor, 1.7 inches in diameter. Repeater amplifiers are placed approximately every 5 nautical miles, and equalizers are placed about every 100 to 150 nautical miles. The electronic repeater amplifiers and equalizers are powered from one end by a 7,000 volt power supply which also feeds over the center conductor. The system operates on two bands: 1.0 to 13.5 MHz and 16.5 to 29.1 MHz.

Digital Carrier

Digital carrier facilities may operate over the same types of media as analog carrier terminals, but all of the voice and control signals are converted to digital data, multiplexed, and sent as a continuous digital data stream over a single channel. *Table 1-5* shows information for digital carrier multiplex systems. Note that the more voice channels handled, the higher the digital signal data rate in bits per second (bps).

Table 1-5. *Digital Carrier and Multiplex Systems*

Medium	Digital Signal Designation	Multiplex Designation	Number of VF Channels	Data Rate (Mbps)
T1 Paired Cable 1A-RDS Radio	DS-1	D-Channel Bank	24	1.544
T1C Paired Cable	DS-1C	M1C	48	3.152
T2 Paired Cable	DS-2	M12	96	6.312
3A-RDS Radio FT-3 Fiber Optic	DS-3	M13	672	44.736
T4M Coaxial Cable	DS-4	M34	4032	274.176
WT4 Wave Guide	DS-4	M34	4032	274.176
DR18 Radio	DS-4	M34	4032	274.176
FT4 Fiber Optic Cable	DS-4	M34	4032	274.176

CCIS Facilities

As previously discussed, the Common Channel Interoffice Signaling system operates over a channel with VF bandwidth that carries all the supervisory control signals for several VF channels carrying voice only. The facilities necessary to accomplish this are shown in *Figure 1.22*.

The CCIS link consists of a terminal access circuit (TAC), a signaling terminal, and a modem (discussed in Chapter 6) at each exchange; and a channel to a control computer called a Signal Transfer Point (STP). Several exchanges may be interconnected with two or more STPs for back-up protection against equipment failure.

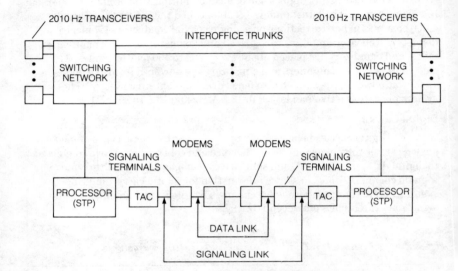

Figure 1-22. *CCIS Signaling Facilities*
(Copyright© 1980, American Telephone and Telegraph Co.)

CCIS is a major improvement in network control. It provides higher signaling speed, much greater information capacity, and two-way signaling even while a conversation is in progress on a trunk. Also, it provides some benefits of separate channel signaling such as freedom from accidental disconnection by voice or extraneous tones that appear to be control tones, protection from misuse of the network by simulating in-band signals, and freedom from mass seizures of whole trunk groups due to carrier or multiplex system failures.

SYSTEM OPERATING CONDITIONS

Telephone systems have certain specifications that are used as guidelines throughout the industry. Here are some of these.

Telephone Set

The telephone set is designed to operate under a wide range of electrical, mechanical, and acoustical conditions. Some of the design parameters are dictated by human factors such as sound pressure levels, handset dimensions, etc.; some are historical carryovers such as ringing voltage and frequency; and some, such as minimum line current for satisfactory carbon transmitter and relay operation, are dictated by physical properties of the materials used in the telephone set. *Table 1-6* lists some operating parameters and limits for subscriber telephone sets in the U.S. and in some Western European countries.

Table 1-6. *Operating Parameters and Limits*

Parameter	Typical U.S. Values	Operating Limits	Typical European Values
Common Battery Voltage	− 48 VDC	− 47 to − 105 VDC	Same
Operating Current	20 to 80 mA	20 to 120 mA	Same
Subscriber Loop Resistance	0 to 1,300 ohms	0 to 3,600 ohms	
Loop Loss	8 dB	17 dB	Same
Distortion	− 50 dB total	N.A.	
Ringing Signal	20 Hz, 90 VRMS	16 to 60 Hz, 40 to 130 VRMS	16 to 50 Hz, 40 to 130 VRMS
Receive Sound Pressure Level	70 to 90 dBspl*	130 dBspl	Varies
Telephone Set Noise		less than 15 dBrnC**	

* dBspl = dB sound pressure level
**dBrnC = dB value of electrical noise referenced to − 90 dBm measured with C message weighting frequency response

TELEPHONE EQUIPMENT REGISTRATION

Up until 1955, the FCC supported the monopoly of the common carriers in the area of terminal equipment and devices attached to terminal equipment. Terminal equipment is the equipment, such as telephone sets, that is attached to the network. In 1955, the FCC ruled that the carriers could not prohibit attachment of devices which were "privately beneficial without being publicly harmful" to equipment leased by the carrier. (This was the so-called "Hush-a-Phone" case.) This policy was then extended in the Carterfone decision in 1968 to apply to equipment connecting to the switched network. In 1975, the FCC determined that any piece of equipment registered with the FCC could be connected directly to the telephone network (this did not include coin telephones and party line telephones).

The rules for registration of equipment to be connected to the network are set out in the Rules and Regulations of the Federal Communications Commission, Part 68: Connection of Terminal Equipment to the Telephone Network, dated July 1977. This document (available from the U.S. Government Printing Office) describes the types of equipment which must be registered, the electrical and mechanical standards to be met, and standard plug and jack arrangements to be used in connecting terminal equipment to the network. It also specifies the procedures to be followed in testing equipment for compliance to Part 68 rules, and the method of applying for certification.

Quiz for Chapter 1

1. The telephone was invented by:
   a. Watson.
   b. Bell.
   c. Strowger.
   d. Edison

2. The central office detects a request for service from a telephone by:
   a. a flow of loop current.
   b. no loop current.
   c. a ringing signal.
   d. dial pulses.

3. Which kinds of signal are transmitted on the local loop?
   a. voice
   b. tones
   c. pulses
   d. all of the above

4. Which office class is the local central office?
   a. 2
   b. 3
   c. 4
   d. 5

5. Which exchange is used to connect between central offices when a direct trunk is not available?
   a. local
   b. tandem
   c. toll
   d. any of the above

6. Which of the following is a type of dc signaling?
   a. loop current
   b. reverse battery
   c. E&M
   d. all of the above

7. The voice frequency channel pass band is:
   a. 0 to 4,000 Hz.
   b. 300 to 3,000 Hz.
   c. 8,140 to 8,188 Hz.
   d. none of the above.

8. What is used to transmit more than one conversation over a path?
   a. hybrid
   b. tandem
   c. multiplexing
   d. all of the above

9. The common channel interoffice signaling method:
   a. uses the same channel for signaling as for the related conversation.
   b. uses a separate channel for signaling only.
   c. carries the signaling for only one related conversation.
   d. is used on local loops.

10. Telephone switching is accomplished by:
   a. manual switchboard.
   b. step-by-step switches.
   c. crossbar switches.
   d. any of the above.

11. The step-by-step switch:
   a. was invented by Strowger.
   b. generates much noise.
   c. cannot operate directly from DTMF tones.
   d. all of the above.

12. Time division multiplexing is used for:
   a. analog transmission.
   b. digital transmission.
   c. both of the above.

The Conventional Telephone Set

ABOUT THIS CHAPTER

In Chapter 1, the telephone set functions were briefly covered in showing how a call was initiated and connected. In this chapter, the conventional telephone set functions will be examined in detail. Then, in Chapters 3 and 4, we'll discuss the electronic circuits that perform these functions in electronic telephone sets.

SWITCHHOOK

On-Hook

Figure 2-1 is a block diagram of a telephone set which shows the major functions. The ringer circuit, which will be discussed later, is always connected across the line so it can signal an incoming call. The remainder of the telephone set is isolated from the line by the open contacts of the switchhook when the handset is on-hook. No dc flows (except possibly a small leakage current) since the ringer has a capacitor that blocks dc flow through it.

Off-Hook

When the handset is lifted off-hook to make a call, the switchhook contacts, S1, S2 of *Figure 2-2a*, close. Loop current flows from the central office battery through the telephone set and through a relay coil at the central office. When sufficient current flows in the relay coil, this relay is energized and its closed contacts signal to other central office equipment that a subscriber telephone is off-hook. A line finder searches until it finds the line with the off-hook signal. The line finder then sets up a connection for the switching equipment to begin receiving the telephone number. At this point, a dial tone generator is connected to the line to signal the caller to proceed with dialing. Dialing may be done by pulsing (interrupting) the loop current or by sending audio tones. When the first dialed digit is received at the central office, the dial tone is removed from the line.

PULSE DIALING

In the conventional telephone set, pulse dialing is accomplished with a rotary dial having ten equally spaced fingerholes as shown in *Figure 2-1*. The number of dial pulses resulting from one operation of the dial is determined by how far the dial is rotated before releasing it. The regularly spaced holes in the dial plate and the finger stop make it easy to rotate the dial the correct amount for each digit. This action winds up a spring which rotates the dial back to the rest position when it is released. A small governor inside the dial causes it to

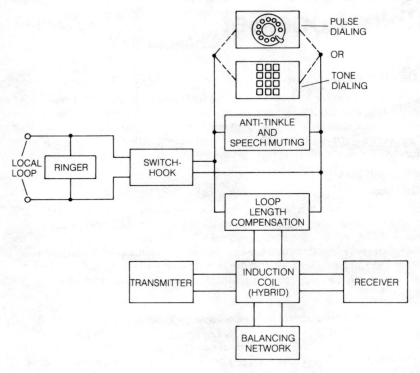

Figure 2-1. *Telephone Set Block Diagram*

return at a constant rate of rotation. A cam turned by the shaft through a gear operates the switch contact S3, shown in *Figure 2-2a*, which opens and closes the local loop circuit during the return rotation of the dial. (The loop is not broken during forward rotation of the dial.) Opening the loop circuit interrupts the loop current flow of 20 to 120 milliamperes and closing the circuit permits the loop current to flow again. Thus, pulse dialing produces a series of current pulses in the loop circuit. One pulse is sent for the digit 1, two for the digit 2, etc., up to ten pulses for the digit 0.

Pulse Timing

Dial pulses were originally conceived to operate electromechanical switching systems. The mechanical inertia associated with such systems set an upper limit on the operating rate of about 10 operations per second. Thus, mechanical rotary telephone dials were designed to produce a nominal rate of ten pulses per second.

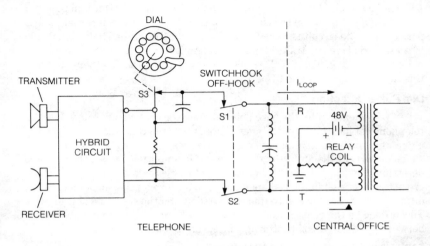

a. Dial Circuit with Switchhook Closed (Off-Hook)

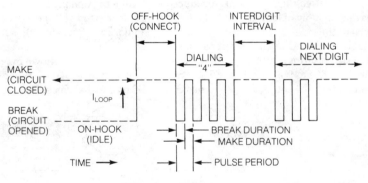

b. Dial Pulse Timing (for "4")

Pulse Period	=	Break Duration + Make Duration (100 Milliseconds Nominal)
Pulse Rate	=	Pulses Per Second = 1000 ÷ Pulse Period (MS)
Percent Break	=	100 × Break Ratio
	=	100 × Break Duration ÷ Pulse Period
Interdigit Interval	=	700 Milliseconds Nominal (may range from 600 to 900 depending on system).

Figure 2-2. *Dial Pulses*

Figure 2-2b shows the timing relations of dial pulses. Note that the number of breaks represent the number dialed. One timed interval of circuit opening and closing, called a dial pulse period, is normally 100 milliseconds long, giving the desired pulse rate of 10 pulses per second. (One second equals

1,000 milliseconds, thus 1,000/100 = 10 pulses per second.) One dial pulse consists of a period when the circuit is open (called the break interval), and a period when the circuit is closed (called the make interval). The nominal value of these periods in the U.S. telephone system is 60 milliseconds break and 40 milliseconds make. This is called a 60 percent break ratio. In other countries, this ratio is usually around 67 percent.

Dial Pulse Detection

For each mile, the wire pair between the central office and the telephone set has shunt capacitance of about 0.07 microfarad, series inductance of about 1.0 millihenry and series resistance of about 42 ohms which distort the dial pulses in duration and in amplitude. Thus, the dial pulse detection circuits at the central office must be able to detect dial pulses that are not perfect rectangular-shaped pulses. The circuits also must decode the difference between successive pulses of a digit and the start of a new digit. This is done on the basis of time. As shown in *Figure 2-2b*, the nominal interval between dialed digits (the interdigit interval) is 700 milliseconds.

ANTI-TINKLE AND SPEECH MUTING

High-voltage spikes are produced each time the dial pulsing contacts interrupt the flow of loop current. These spikes of increased voltage can cause the bell of the ringer to sound as the pulses are generated. The ringing is fairly soft, like a tinkle, thus the circuit to prevent it is called an anti-tinkle circuit. This circuit is often combined with the speech muting circuit as shown in *Figure 2-3*. The extra switches, S5 and S6, are part of a normal dial. When the dial is rotated, the contacts of switches S5 and S6 on the dial close. This shorts

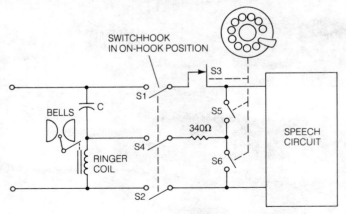

S5 and S6 are mechanically coupled to the dial. They close and remain closed when the dial is rotated. They open when the dial returns to its rest position.

***Figure 2-3.** Anti-Tinkle and Speech Muting*

the speech circuit to prevent loud clicks in the receiver and possible damage to the speech circuit. The closed contacts also shunt the ringer coil with a 340 ohm resistor to prevent bell tinkle. In this circuit, the ringer capacitor C also serves as a spark quencher to suppress arcing at the dial pulsing contacts.

TONE DIALING

Some telephone sets use the method called dual tone multifrequency (DTMF) for sending a telephone number. These can be used only if the central office is equipped to process the tones. As shown in *Figure 2-4*, instead of a rotary dial, these telephone sets are equipped with a push-button keypad with 12 keys which represent the numbers 0 through 9 and the symbols * and #. (As shown dotted in *Figure 2-4*, some special purpose telephones have a fourth column of keys for a total of 16 keys.) Pressing one of the keys causes an electronic circuit to generate two tones in the voice band. There is a low frequency tone for each row and a high frequency tone for each column. Pressing key 5, for example, generates a 770 Hz tone and a 1,336 Hz tone. By using the dual tone method, 12 unique combinations are produced from only seven tones when the 12 position keypad is used.

The frequencies and the keypad layout have been internationally standardized, but the tolerances on individual frequencies may vary in different countries. The North American standard is ± 1.5% for the generator and ± 2% for the digit receiver.

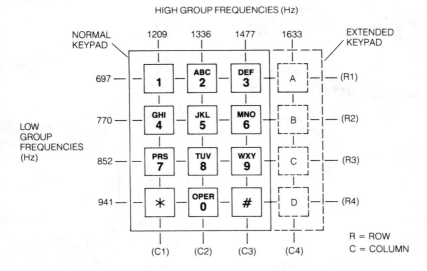

Figure 2-4. DTMF Keypad and Frequencies

Tone Generation

A typical DTMF circuit is shown in *Figure 2-5*. The switches S1, S2, and S3 are shown in the inactive position. With the switchhook in the off-hook position, loop current flows through RV1, L1A, L2A, and via the hybrid to the line. Transistor Q1 is off. Capacitors C1 and C2 are disconnected at one end by the open contacts of S1 and S2.

When a key is pressed, mechanical couplings, called the row rod and column rod, for that key position close the appropriate S1 and S2 contacts to connect C1 to a tap on L1A and C2 to a tap on L2A. This establishes the resonant circuits for the required low-group tone (L1A-C1) and high-group tone (L2A-C2).

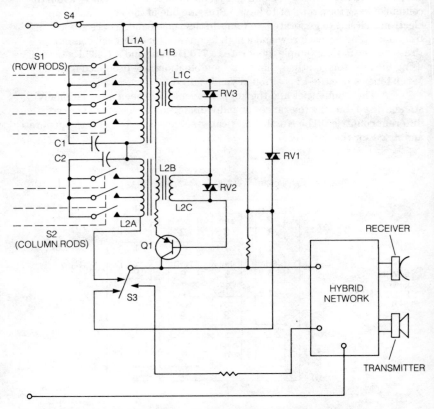

Switches S1, S2 and S3 are shown in their inactive position. They change position when a key is pressed. S4 is activated by the switchhook. It is shown in the off-hook position.

Figure 2-5. *DTMF Circuit*

The mechanical arrangement of the pushbutton and switches is such that the resonant circuit connections just described are made with only partial depression of the button. At this point, S3 is still in the position shown. Further depression of the button causes S3 to change position. This interrupts the dc flow through L1A and L2A and shock-excites the two resonant circuits into oscillation. At the same time, S3 connects battery voltage from the line to the collector of Q1. The transformer coupling between L1A, L1B and L1C, and between L2A, L2B and L2C cause Q1 to sustain the oscillations and modulate the loop current to transmit the two tones to the central office. The transmitter and receiver are shunted when S3 switches, but the outgoing tones can still be heard in the receiver at a low level.

Tone Detection

The tones used have been carefully selected so that the processing circuits (called the digit receiver) in the central office will not confuse them with other tones that may occur on the line. The digit receiver has frequency selective filters that pass only the frequencies used for DTMF. It also has timing circuits which make sure a tone is present for a specified minimum time (about 50 milliseconds in North America) before it is accepted as a valid DTMF tone.

After a connection to an answered phone has been made, the digit receiver is out of the circuit and the DTMF tones can be transmitted the same as speech. This permits the use of DTMF tones as data communications for entering orders at a remote terminal or obtaining information from a remote data base.

Time Comparison

DTMF dialing is much faster in principle and in practice than dial pulsing. Using DTMF, the time required to recognize any digit tone is only 50 milliseconds with an interdigit interval of another 50 milliseconds. Thus, the total time to send *any digit* is about 100 milliseconds.

In contrast, dial pulsing requires 60 milliseconds break and 40 milliseconds make for each dial pulse for a total time of 100 milliseconds per *dial pulse*. Thus, for dial pulsing, each higher digit number requires more time because the number of pulses per digit increases. Also, the interdigit interval of about 700 milliseconds for dial pulsing is much longer. By using the number 555-555-5555, an average time for pulse dialing a long distance call can be obtained as follows:

5 pulses per digit $\times$ 100 msec per pulse $\times$ 10 digits = 5 seconds
Interdigit interval $\times$ (number of digits $-$ 1) = 700 ms $\times$ 9 = 6.3 seconds
Total time for dial pulsing = 5 + 6.3 = 11.3 seconds

DTMF dialing for the same number takes:

Number of digits $\times$ 100 ms per digit = 10 $\times$ 100 ms = 1 second.

These times are the minimum. Actual physical operation of the dial and keypad add to these times, but the time savings from the use of DTMF are substantial when summed for the many telephone calls made in a day. At the local exchange, the digit receiver and memory required to hold the digits as they are keyed are shared among many incoming lines. A reduction in the average time per call that each is in use (called the holding time) means fewer digit receivers are required for the same service. The end result is that less capital equipment is required for a given number of lines; therefore, the cost of the exchange is reduced.

Coupling the DTMF Generator to the Line

The requirements for the proper interfacing of the DTMF generator to the line are:

1. The correct DC voltages and loop currents must be maintained on any loop length.
2. The tones must have proper amplitude and distortion characteristics.
3. The DTMF generator must have the proper impedance to match the line.

Power

Problems may arise when powering the DTMF circuits from the line in either of the two extreme cases; long loops or short loops.

Long loops reduce the amount of current and voltage available for electronic circuits in the telephone; thus, the tone dialing circuits need to operate from a supply voltage as low as 3 volts.

The minimum working dc voltage ($V_{DC(MIN)}$) for the DTMF generator and interface circuit will be the sum of the peak voltages of the two tones ($V_{LPK} + V_{HPK}$), plus the desired regulated voltage (V_{Reg}), plus the voltage drop necessary to achieve the regulating voltage ($V_{BE} + V_{CE(SAT)}$); e.g.;

$$V_{DC(MIN)} = (V_{LPK} + V_{HPK}) + V_{Reg} + (V_{BE}) + V_{CE(SAT)})$$
$$= 1.24V + 3V + 1.2V$$
$$= 5.44V$$

Short loops require the telephone set to sink (absorb) high loop current or handle high dc voltage if no provision is made at the central office to regulate or limit the current and voltage supplied to the loop. In either case, the interface must maintain proper dc voltage regulation if it is to supply voltage to other circuits.

Level

The sending levels of the DTMF tones are referenced to 0 dBm (1 milliwatt of power dissipated in a 600 ohm impedance). The high frequencies have a level 2 dB above the level of the low frequencies to compensate for losses in transmission. *Figure 2-6* is the curve usually shown to specify the sending levels of the tone pairs.

HIGH FREQUENCY LEVEL (dBm)

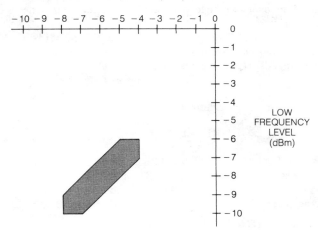

Figure 2-6. *DTMF Sending Level Specification*

Distortion

Allowable distortion of the transmitted tones is specified in
several ways:

1. The total power of all unwanted frequencies shall be at least 20 dB below the
   level of the frequency pair which has the lowest level.
2. The level of unwanted frequencies produced by the pair shall be:
   a. Not more than -33 dBm in the band from 300 to 3,400 Hz.
   b. Not more than -33 dBm at 3,400 Hz and falling 12 dB per octave
      to 50 kHz.
   c. Not above -80 dB above 50 kHz.
3. Distortion in dB is defined as:

$$\text{Distortion} = 20 \, \text{Log}_{10} \, \frac{\sqrt{V_1^2 + V_2^2 + \text{---} + V_N^2}}{\sqrt{V_L^2 + V_H^2}}$$

Where: V_1 through V_n are the unwanted frequency components
V_L is the low frequency tone
V_H is the high frequency tone

Impedance

When the DTMF generator is active, it must present the correct dynamic impedance to the loop. A nominal 900 ohm impedance is the normal requirement.

When the DTMF generator is inactive, it must present a low dynamic impedance when connected in series with the speech circuit and a high dynamic impedance when connected in parallel.

Return Loss

Return Loss is defined as:

$$RL = 20 \, \text{Log}_{10} \, \frac{Z_L + Zg}{Z_L - Zg}$$

where Z_L is the line impedance and Zg is the output impedance of the telephone set generating the signal. RL must be greater than 14 dB in the frequency band from 300 to 3,400 Hz. RL must be greater than 10 dB in the frequency bands from 50 to 300 Hz and from 2,400 to 20,000 Hz.

Advantages of DTMF

In summary, DTMF dialing is replacing pulse dialing because it:

1. decreases dialing time,
2. uses solid-state electronic circuits,
3. can be used for end-to-end signaling after the call is connected (low-speed data transmission),
4. reduces local exchange equipment requirements,
5. is more compatible with electronically (stored program) controlled exchanges.

TRANSMITTER

The part of the telephone into which a person talks is called the transmitter. It converts speech (acoustical energy) into variations in an electric current (electrical energy) that can be transmitted through the transmission system to the receiver of the called telephone. The most common telephone transmitter in use today is in principle like the one invented about 100 years ago by Thomas A. Edison.

Construction

As shown in *Figure 2-7a*, it consists of a small, two-piece capsule filled with thousands of carbon granules. The front and back are metallic conductors and are insulated from each other. One side of the capsule is held fixed by a support that is part of the handset housing. The other side is attached to a diaphragm which vibrates in response to the air pressure variations caused by speaking into it. The vibrations of the diaphragm vary the pressure on the carbon granules. If the granules are forced together more tightly, the electrical resistance across the capsule decreases. Conversely, if the pressure on the granules is reduced, they move apart and the resistance increases. The current flowing through the transmitter capsule varies because of the varying resistance; thus, the varying air pressure representing speech is converted to a varying electrical signal for transmission to the called party. Other carbon transmitters may be constructed differently, but operate the same way.

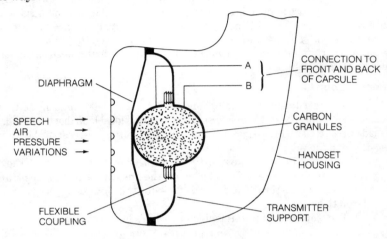

a. Construction
(Reprinted with permission, abc TeleTraining, Inc., Geneva, Ill.)

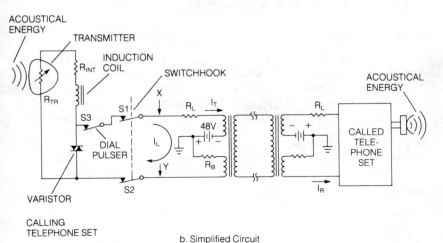

b. Simplified Circuit

Figure 2-7. *Telephone Transmitter*

Operation

To understand how the level of the loop current varies with the speech level, refer to *Figure 2-7b*. The carbon granule transmitter is represented in the circuit by the variable resistance R_{TR} which changes as a person speaks. The voltage applied across the transmitter is constant; therefore, as the sound energy varies the resistance R_{TR} of the transmitter, the current in the circuit varies in the same manner as the sound intensity. Because the voltage applied is another energy source, the current variations in the circuit deliver electrical energy to the called receiver that produces a greater amount of acoustic energy than that which caused the original resistance variations at the transmitter. Thus, the original speech signal has been amplified.

Effect of Loop Length

The resistance of the telephone set is typically 400 ohms in the U.S. if measured at points X and Y in *Figure 2-7b* with the local loop disconnected. It is made up of R_{TR} and R_{INT}, where R_{INT} lumps together the rest of the resistance in the telephone besides R_{TR}.

R_L represents the line resistance of both wires. It varies with the length of line to the central office; the longer the line, the larger the resistance. R_B is a balancing resistor in the central office; for this discussion, R_B is considered to be zero.

The loop current is determined by Ohm's law as follows:

$$I_L = \frac{\text{Central Office Voltage}}{R_{TR} + R_{INT} + R_L}$$

The central office voltage in the U.S. is usually 48 volts, therefore:

$$I_L = \frac{48V}{R_{TR} + R_{INT} + R_L}$$

As the total resistance in the local loop $(R_{TR} + R_{INT} + R_L)$ gets larger as the line length from the telephone set to the central office increases, the current gets smaller. Also, as the total resistance gets larger, the variations in R_{TR} as a percentage of the total resistance become much smaller. Therefore, for the same speech input, the variations in the local loop current become smaller as the line length increases. This causes the speech level to go down at the called telephone's receiver. This is not a desirable condition and methods to provide automatic compensation for line length have been developed.

Compensation for Loop Length

For ease and uniformity of telephone use, it is desirable that the speech on all calls arrive at the exchange at about the same volume or level regardless of loop length. To achieve this, modern telephone sets have automatic compensation circuits. In *Figure 2-7b*, the varistor resistance decreases as loop current increases with shorter loops. This resistance bypasses some of the current around the transmitter so that the transmitter current is about the same as it is on a long loop. Thus, the varistor automatically adjusts the speech level so that a relatively constant speech level will appear at the exchange regardless of the distance to the telephone set.

Distortion

The principal function of the transmitter is to produce an electrical output waveform that is shaped like the input speech waveform. This output is called an analog signal since the transmitter's electrical output is an analog of (similar to) the acoustic input. Any difference between the waveforms is called distortion. Some forms of distortion are readily apparent to the listener, others are not. The most obvious distortion in the telephone system is the "flattening" quality given to the voice. Some of the distortion is caused by the carbon transmitter and, as shown in Chapter 1, some is due to the fact that some frequencies contained in human speech are not transmitted. This doesn't impair the intelligibility of the speech, but a listener can quickly recognize a recording of a telephone conversation when compared to a recording of someone speaking through a high fidelity sound system. Much of the distortion caused by the carbon transmitter can be eliminated in electronic telephones that have an active speech amplifier by using a low distortion microphone.

Microphones for Use as Transmitters

All microphones may be defined as electro-acoustic transducers, which convert the varying pressure of a sound wave to a varying current or voltage by means of a mechanical system. Although there are many types of microphones, only two besides the carbon type combine the ruggedness and low cost required for application in telephone sets. One type is the electromagnetic or electrodynamic microphone and the other type is the electret microphone.

Electrodynamic Microphone

This type has a diaphragm attached to a wire, coil, or ribbon in a permanent magnetic field as shown in *Figure 2-8*. Sound waves striking the diaphragm cause the coil to move. This movement in the presence of a permanent magnetic field induces a current in the coil proportional to the movement; thus, the sound is converted to a varying electric current which can be amplified and transmitted.

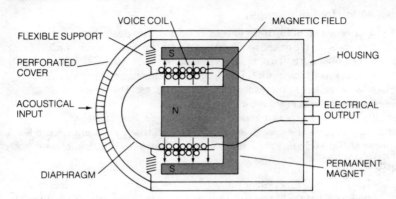

Figure 2-8. *Electrodynamic Microphone (Cross Section)*

Electret Microphone

This type has been known for many years, although its practical application to telephony awaited development of better dielectric materials and low-cost amplifiers. An electret is the electrostatic equivalent of a permanent magnet. It is a dielectric which will hold an electric charge almost indefinitely.

The relationship of voltage (V), capacitance (C) and charge (Q) is given by the following equation:

$$V = \frac{Q}{C}$$

When an electret is the dielectric between two metal plates, it forms a special kind of capacitor. The charge Q is held permanently in the electret material. Therefore, if one of the plates of the capacitor is the diaphragm of the microphone, the voltage at the terminals will vary according to the diaphragm movement. The variations are very small and must be amplified.

The dielectric material of a modern electret microphone is a fluorocarbon material. It is produced as a foil metallized on one side. The material is charged by placing it in a corona discharge and forcing the electrons into the foil with an external electric field. The resulting film can be assembled into a metal can of about 18mm (0.7 in.) diameter and 18mm length.

One type of construction is shown in *Figure 2-9*. The resulting microphone is similar to the studio ribbon-type capacitor microphone, but does not need a polarizing voltage. Like the ribbon microphone, the device has high internal impedance. Matching to a low-impedance circuit is sometimes done with a source follower FET (field-effect transistor) mounted inside the microphone capsule. The diaphragm has a working diameter of 15mm (0.59 in.) and a capacitance of 15 picofarads. The stability of the charge on the foil is acceptable for commercial use and the total harmonic distortion of the device can be less than 1% compared with 8 to 10% for a carbon granule microphone.

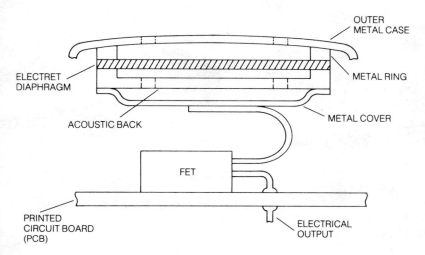

Figure 2-9. *Electret Microphone Assembly*

RECEIVER

The receiver converts the varying electrical current representing the transmitted speech signal (I_T in *Figure 2-7*) to variations in air pressure useable by the human ear. A typical electromagnetic receiver, shown in *Figure 2-10a*, consists of coils of many turns of fine wire wound on permanently magnetized soft iron cores that drive an armature. The armature is a diaphragm made of a soft iron material.

A key requirement for an electromagnetic receiver is a permanent magnet to provide a constant bias field for the varying electromagnetic field to work against; otherwise, both positive and negative currents would push the armature in the same direction. The varying electrical current representing speech (I_R in *Figure 2-7)* flows through the coils and produces a varying electromagnetic field. It alternately aids and opposes the permanent magnet field; thus, it alternately increases and decreases the total magnetic field acting on the diaphragm as shown in *Figure 2-10b*. This causes the diaphragm to vibrate in step with the varying current and moves the air to reproduce the original speech that caused the current changes. (The principle of a permanent magnet providing a bias field also is used in loudspeakers for audio reproduction in many kinds of electronic equipment).

The receiver shown in *Figure 2-10c* operates similarly except the armature is a separate part and is connected to a conical non-magnetic diaphragm. The rocking action of the armature causes the aluminum diaphragm to vibrate to reproduce the original speech.

It is interesting to note that the electromagnetic receiver was a central element of Alexander Bell's original telephone patent.

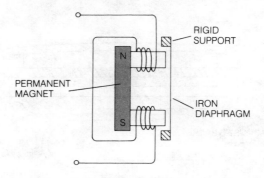

a. Standard Receiver

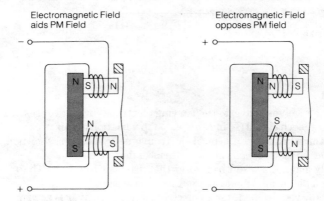

b. Reaction of Magnetic Fields Produce Movement of Armature

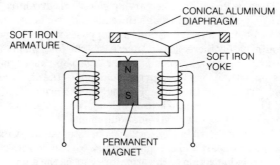

c. Rocking Armature Receiver

Figure 2-10. *Telephone Receivers*

ELECTROMECHANICAL RINGER

When a call has been connected through to the central office serving the called subscriber's local loop, the central office must send a signal to the called party that a call is waiting to be answered. This signaling function is called ringing.

From the telephone company's point of view, it is desirable that a called subscriber answer an incoming call as quickly as possible. While the telephone is ringing, expensive common control equipment in the exchanges, and for toll calls, even more expensive transmission facilities are tied up. When in this condition, these facilities are generating no income. A call arrival alerting system, therefore, must signal incoming calls in an urgent manner and must be loud enough to be heard at considerable distance. For residential phones, it is usually desirable to hear the signal from outside the house. The traditional metallic gong bell struck by a metal clapper fulfills these requirements. It was patented in 1878 by Thomas A. Watson (Mr. Bell's assistant).

Operation

As shown in *Figure 2-11a*, the armature is pivoted in the middle and two electromagnets, formed by coils wound on permanently magnetized iron cores, alternately attract and repel opposite ends of the armature in response to an applied alternating current. The armature drives a hammer to alternately strike two bells or gongs.

The coils are wound so that the magnetization in the coils due to an applied electric current is of opposite polarity on each of the outer poles of the E-shaped structure. Like the receiver structure, the permanent magnet provides a bias on the armature so that it will be attracted first to one side, then the other, as the applied ac sets up an alternating magnetic field. The permanent magnet also accelerates the hammer as the armature closes so that the hammer strikes the bell harder to produce a louder sound. A mechanical adjustment to regulate the loudness is often included in the telephone set.

A simplified circuit is shown in *Figure 2-11b*. V_R is the ringing voltage applied from the central office. In the U.S., it is usually about 90 VRMS at a frequency of 16 to 60 Hz. It produces the alternating current to drive the ringer. The capacitor C passes the ac for ringing, but blocks the flow of any dc from the local loop battery. The value of inductance for the coils and the value of capacitance are chosen so that the ringer circuit presents a high impedance to the voice frequencies. Since the handset is on-hook, the switchhook is open so the ringing voltage is not applied to any other telephone circuits.

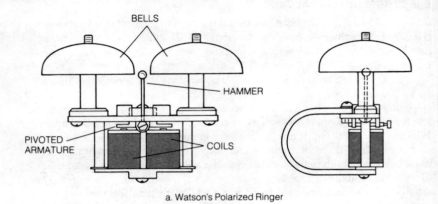

a. Watson's Polarized Ringer

b. Ringer Circuit

Figure 2-11. Telephone Ringer

Ringing Generator

The voltage source for the alternating current to operate the ringer was at first supplied by a hand-cranked magneto, also patented by Mr. Watson in 1878. The magneto produced about 75 volts at about 17 Hz. The high voltage was necessary due to the inefficiency of existing magnetic materials and system transmission lines. Although the ringer in modern phones is more efficient because of improvement in materials and manufacturing methods, the high voltage still is used today.

In most central offices, the ac ringing signal is generated by a dc motor driving an ac generator or by a solid-state inverter. The inverter produces an ac voltage from a dc voltage without any moving parts. Both are called ringing machines and are powered by the –48 Vdc central office supply. Since many lines may need ringing at the same time, several ringing machines are used to distribute the load. Usually, these machines are operated so that the ringing times between machines is staggered. By doing this, at least one of the machines will be in the ringing portion (rather than the silent portion) of the cycle at any time. Thus, any new call will have an active ringing signal available immediately rather than having to wait until a silent period ends.

The ac signal is applied to the loop in timed ON and OFF intervals to produce a ringing cadence. *Figure 2-12* shows two typical cadences. The first is widely used in the U.S. and Europe, and the second is used in the U.K. In some private systems (PABX's), different cadences are used to indicate whether the source of a call is internal or external.

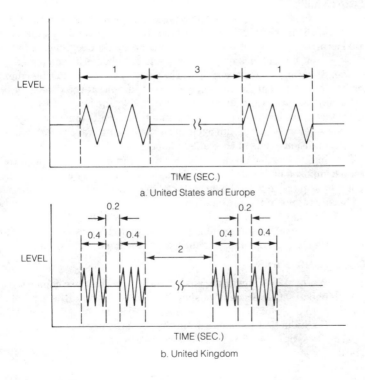

Figure 2-12. Ringing Cadence

Loop Current Detection in the Presence of Ringing

When the telephone is being rung, an alternating current flows in the loop, but no dc flows. If the call is answered while the bell is ringing, it is necessary for the exchange to detect that direct current has begun to flow at the same time that ringing current is flowing, so that the ringing can be removed before the receiver is applied to the called party's ear (this time period is normally considered to be 200 milliseconds). This process is called "ring trip".

The value of the current in the local loop varies widely. It depends on the exchange battery voltage, the current limiting resistance inserted in the loop at the exchange to limit short circuit current (ranges from 350 to 800 ohms), the resistance of the local loop (which may vary from 0 to about 1900 ohms), and the resistance of the telelphone set itself (between 100 and 400 ohms).

The circuits in the central office can detect loop direct currents in the range of 6 to 25 milliamperes as indicating an off-hook condition. However, the detection circuit must not be so sensitive that leakage currents between the tip and ring sides of the loop, or leakage to ground from either conductor, is interpreted as an off-hook condition. The problem is further complicated by the fact that a typical residence may have several telephones all connected to the same line. Since all phones are rung at the same time, the resulting ringing current may reach 50 mA ac, while the loop direct current at the time the call is answered may be lower than 20 mA dc. Since during one half-cycle of the ac wave the central office can't tell dc from ac, ring trip circuits must average the sensed current over at least one full cycle of the ac ringing wave to determine if an off-hook condition exists at the called phone.

The ringing method discussed above has ringing current applied to the loop superimposed on the dc battery voltage. Another way is to apply the ringing current as pure ac with the dc restored in between bursts of ringing current. The advantage of the first method is that the answer condition can be detected even during the ringing cycle by observing the average dc level. Detecting answer during ringing when only ac is applied is more difficult. This requires detection of the impedance change when the handset is lifted.

THE HYBRID FUNCTION

Figure 2-13 shows how the transmitter and receiver are connected within the telephone set. It also shows simplified connections at the central office exchange. In particular, note the components labeled "induction coil" and "hybrid". The induction coil name is another of the carryover terms from the early days, but one of its functions in the circuit is similar to that of the hybrid.

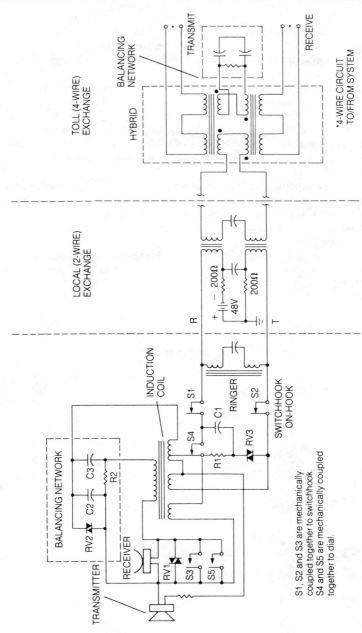

Figure 2-13. The Hybrid Function in the Telephone and Exchange

The function of a hybrid is to interface a 2-wire circuit to a 4-wire circuit to permit full-duplex operation. In communications, full-duplex means that transmission of signals over the circuit can occur in both directions at the same time. Two-wire circuits are used for the millions of subscriber local loops because they are cheaper. Four-wire circuits, which have two wires for each direction, are used for almost all other circuits in the network. Physical separation of the transmit and receive signals is needed on these circuits so that all types of transmissions can be handled—digital data as well as voice—and so that electronic amplifiers can be used in the transmission path to increase the signal level. A hybrid is used at the central office to interface the local loop to trunks and between trunks at some older offices where only 2-wire switching is available.

The telephone handset also is a 4-wire circuit with two wires used for the transmitter connection and two for the receiver connection. This is shown in *Figure 2-11* and *2-13*. It can be represented by the circuit as shown in *Figure 2-14*; however, the telephone circuit really evolved from an anti-sidetone circuit rather than a 2-wire to 4-wire hybrid. For this reason, the circuit in the telephone set more commonly is called an induction coil and the interconnections of *Figure 2-13* are somewhat more complicated than the 2- to 4-wire hybrid of *Figure 2-14*.

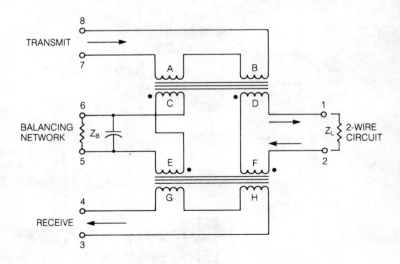

Figure 2-14. *Hybrid*

Operation of the Hybrid

The hybrid is a multiple winding transformer. Actually, as indicated by the diagram in *Figure 2-14*, it usually consists of two interconnected transformers in one physical container. Like any transformer, it uses electromagnetic coupling, i.e., a changing current in winding A produces a changing magnetic field, which in turn induces a changing voltage in winding C on the same core. When the winding C is connected in a circuit, the changing voltage causes a changing current to flow in the circuit. Thus, the changing current pattern or waveform of the input winding (primary) is reproduced in the output winding (secondary) without physical connection between the two circuits. The ratio of the input voltage to the output voltage and the input current to the output current depends on the ratio of the number of turns in the primary winding to the number of turns in the secondary winding. The turns ratio also allows impedance matching between coupled circuits. The impedance in the secondary appears in the secondary circuit as $N^2 \times Z_p$, where N is the ratio of the secondary turns to the primary turns and Z_p is the primary impedance. Impedance matching is important in circuits to reduce power loss and signal reflections.

In *Figure 2-14*, a transmit signal applied to terminals 7-8 is inductively coupled to windings C and D by the current flow through windings A and B. The voltage induced in winding D causes a current to flow through the 2-wire circuit connected to terminals 1-2 for further transmission of the signal. This same current also flows through winding F which induces a voltage in winding H.

In like fashion, the voltage induced in winding C causes a current to flow through the balance network and winding E. The impedance Z_B of the balance network matches the line impedance Z_L. Since Z_B equals Z_L, the turns on coil C and D are equal, and the turns on coil E and F are equal, the same current flows in the circuit of C and E and the circuit of D and F. As a result, the same voltage is induced in windings G and H, since they also have turns equal to each other.

Windings C and E have their connections reversed from the connection of windings D and F. Therefore, the voltage induced in winding G is opposite in phase to the voltage induced in winding H and opposes the voltage in winding H. Since they are equal and opposite, they cancel each other. The net result is that the signal from the transmitter that appears at terminals 7 and 8 is transmitted to terminals 1 and 2, but does not appear at terminals 3 and 4 that go to the receiver. (The signal level at terminals 1-2 is only one-half the level input to terminals 7-8 because the other one-half is dissipated in the balance network.)

A similar effect occurs when a signal appears at terminals 1 and 2 from the line. The current flowing in windings D and F induces a voltage in windings B and H. The voltage in H causes a current to flow through the circuit of the receiver, winding G and winding H. The same current that flows through H flows through G. The current flowing in winding G induces a voltage in winding E and causes a current to flow in the balance network. This current flows through winding C and induces a voltage in winding A. The voltage in winding A is 180° out of phase from that induced in winding B by the current flowing in winding D. This again is due to the reversed connection of windings C and E. Therefore, the winding A voltage and the winding B voltage oppose each other and cancel each other. None of the signal that appears at terminals 1 and 2 from the line appears at terminals 7 and 8, the transmitter terminals.

Sidetone

For the hybrid used in the central office exchange, the balancing network is adjusted so that no transmit signal appears at the receive terminals and no receive signal appears at the transmit terminals. However, for the induction coil arrangement in the telephone set, the balancing network is intentionally unbalanced slightly so that a small amount of the transmitted signal also is fed to the receiver of the talking phone. This signal is called the sidetone.

Sidetone is necessary so that the person can hear his/her own voice from the receiver to determine how loudly to speak. The sidetone must be at the proper level because too much sidetone will cause the person to speak too softly for good reception by the called party. Conversely, too little sidetone will cause the person to speak so loudly that it may sound like a yell at the other end. In the circuit shown in *Figure 2-13*, the values chosen for R2, C2 and C3 and the turns ratio of the windings on the induction coil produce the proper level sidetone for most conditions. The varistor, RV2, provides automatic adjustment of the sidetone level for telephone set installations as variations in loop current occur due to the different length of local loops.

THE ELECTRONIC TELEPHONE

A study of *Figure 2-13* reveals an interesting fact; the conventional dial telephone set contains no "active" circuit components (an active electronic element is one which accomplishes a gain or switching action). In other words, the ordinary telephone performs no amplification of the incoming electrical signals. Whatever electrical signal current level is produced at the central office is what is transmitted down the local loop and is what the listener hears. In fact, the signal level at the listener's receiver is weaker because of line loss in the local loop.

Any amplification of the signal occurs in the central office or in the exchanges and repeaters between central offices. This is done to centralize the expensive and complex equipment for maintenance and so the equipment can be shared by many users. As a result, the local loop and central office have become a type of "centralized processing" system; while the telephone set itself has remained a relatively inexpensive "dumb" device that serves only as an interface and input/output device. Said another way, all the "intelligence" is located in the exchanges. However, this is changing and, as will be shown in this book, the telephone set is evolving into a "distributed processing" system, where there is intelligence not only in the central office, but also in the relatively "smart" telephone sets.

Equivalent Circuits

As the discussions in the following chapters show how the telephone set is evolving and will evolve into its position in the distributed system, equivalent circuits of the telephone system will be useful. *Figure 2-15* shows such a circuit. The exchange, line and telephone set are replaced with equivalent circuits. The series resistance and shunt capacitance of the line is represented by lumped values of resistance R_L and capacitance C_L. The

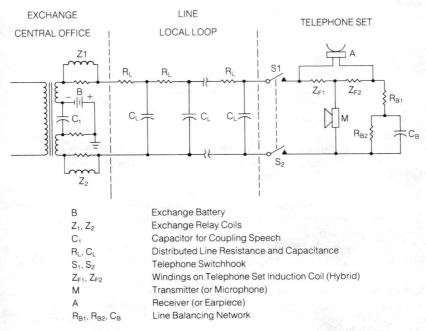

B	Exchange Battery
Z_1, Z_2	Exchange Relay Coils
C_1	Capacitor for Coupling Speech
R_L, C_L	Distributed Line Resistance and Capacitance
S_1, S_2	Telephone Switchhook
Z_{F1}, Z_{F2}	Windings on Telephone Set Induction Coil (Hybrid)
M	Transmitter (or Microphone)
A	Receiver (or Earpiece)
R_{B1}, R_{B2}, C_B	Line Balancing Network

***Figure 2-15.** Equivalent Circuit of Exchange, Telephone Line and Telephone Set*

telephone set is represented by the transmitter, receiver, and equivalent impedances of the induction coil Z_{F1} and Z_{F2}. The balancing network made up of R_{B1}, R_{B2} and C_B is adjusted to match an average line impedance and to provide the correct amount of sidetone.

It is useful to understand the equivalent circuit of the telephone set under the varying conditions of transmission and reception. *Figure 2-16* shows the equivalent circuit when speaking into the transmitter. The voltage generated as a result of speaking is v_T. As it is generated, voltage drops equal to it exist in each of the loops containing Z_L and Z_B. v_o is the voltage that is transmitted down the telephone line as the output resulting from the speech voltage, v_T. The remainder of the v_T voltage, v_{F1} is dropped across the impedance Z_{F1}. In the balance network loop, v_{ZB} is the drop across the balancing network and v_{F2} is the drop across Z_{F2}. The voltages v_{F1} and v_{F2} are equal and oppose each other; therefore, there is no voltage across the receiver because v_{F1} and v_{F2} cancel each other.

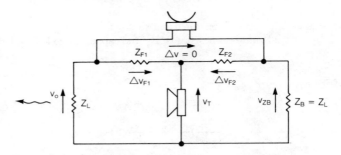

Figure 2-16. *Telephone Set Hybrid—Transmission Equivalent Circuit*

Figure 2-17 shows the equivalent circuit when a signal is being received from the line. The signal is represented as a voltage v_r in series with the line impedance, Z_L. v_r produces voltage drops around the network of v_{F1}, v_T, v_{F2} and v_B. Now v_{F1} and v_{F2} are in the same direction, so that there is a net voltage across the receiver of v_R which causes the receiver to produce sound. The voltage across the transmitter, v_T, is just dissipated in the transmitter and causes no reaction.

These equivalent circuits will be useful in analyzing the use of electronic circuits to replace the passive functions present in the conventional telephone set. For example, the purpose of the present configuration of transmitter, hybrid, and receiver is to transfer as much as possible of the incoming audio frequency power to the receiver, and to transmit as much as possible of the speech power into the transmitter to the rest of the network. If the dc supply from the exchange can be used to power a low-cost, low-voltage audio frequency amplifier in the telephone set, then both the electrical circuit configuration and the physical telephone set can change quite drastically.

The conventional hybrid circuit with its audio transformer can be replaced by a smaller and less expensive circuit using resistive bridge techniques, with the amplifier gain compensating for the losses in the bridge. Since the gain of the amplifier can be controlled, modern low distortion microphones can be used and compensation for loop length can be closely controlled. We'll begin the look at such circuits in the next chapter.

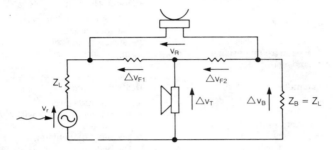

Figure 2-17. *Telephone Set Hybrid—Reception Equivalent Circuit*

WHAT HAVE WE LEARNED?

1. The conventional telephone set performs the following functions: signals the system that it wants the use of the system, sends the dialed telephone number, transmits speech, receives speech and call progress tones, rings to indicate an incoming call, and signals the system to disconnect when the handset is hung up.
2. The transmitter converts speech into electrical signals when its carbon granule resistance is varied by sound pressure on its diaphragm.
3. The receiver converts electrical signals into sound by applying variations in an electromagnetic field produced by the varying electric current to the receiver diaphragm.
4. The dialing of a called telephone number is equivalent to giving the telephone system an address.
5. Voltages as large as 130 VRMS are applied to the telephone set from the line by a ringing voltage generator in the local exchange.
6. Off-hook signal occurs when the handset is off its cradle or hanger and allows current to flow continuously in the local loop.
7. Tone dialing is much faster than pulse dialing and the DTMF tones may be used for low-speed data communications.
8. Muting circuits are used to prevent dial pulse transient voltages from ringing the bell and to protect the speech circuits.
9. The length of the local loop affects the speech signal level so automatic compensation for loop current is provided in the telephone set.
10. A hybrid is a transformer used to interface a four-wire circuit to a two-wire circuit. The induction coil in a telephone set performs this function.

Quiz for Chapter 2

1. The nominal value of the dial pulse break interval in milliseconds in the U.S. is:
   a. 40
   b. 60
   c. 100
   d. 67

2. What type of transmitter is most commonly used in a conventional telephone handset?
   a. carbon
   b. electromagnetic
   c. electret
   d. ceramic

3. Which component in the telephone set has the primary function of compensating for the local loop length?
   a. resistor
   b. varistor
   c. capacitor
   d. induction coil

4. What type of receiver is most commonly used in a conventional telephone handset?
   a. carbon
   b. electromagnetic
   c. electret
   d. ceramic

5. Which component in the telephone set has the primary function of interfacing the handset to the local loop?
   a. resistor
   b. varistor
   c. capacitor
   d. induction coil

6. How many unique tones are used for the 12-key dual-tone multifrequency keypad?
   a. 2
   b. 3
   c. 7
   d. 12

7. Tone dialing takes _____ time than pulse dialing.
   a. less
   b. more
   c. about the same

8. Which of the following are important for proper interface of a DTMF generator to the telephone line?
   a. impedance
   b. tone amplitude
   c. loop current
   d. all of the above

9. The anti-tinkle circuit:
   a. prevents tampering with the telephone.
   b. prevents dial pulsing from ringing the bell.
   c. prevents speech signals from ringing the bell.
   d. all of the above.

10. The sidetone is:
   a. a type of feedback.
   b. determined by the balancing network.
   c. permits the talker to hear his/her own voice.
   d. all of the above.

Electronic Speech Circuits

ABOUT THIS CHAPTER

With this chapter, the discussion moves from a description of the telephone system as it presently exists to a discussion of how it will change and improve as mechanical and conventional electrical devices are replaced by electronic devices, usually in the form of integrated circuits. These devices have most of the required components on the chip, with connections provided for outboard components such as resistors and capacitors which are used to "program" the chip; that is, to set the electrical operating parameters for specific applications. In this way, one design can be used for a variety of similar, yet different requirements. This approach lowers cost and improves performance. In this chapter, the emphasis is on the circuits that provide two-way speech in the telephone set.

But first an important point must be made. Any improvement to any part of the telephone system, whether it be in the telephone set, the switching offices, or the transmission lines, must be not only better, more reliable, less expensive, easier to manufacture, and all the other features associated with progress; but also it must, above all, be compatible with the existing system. If it were possible to start all over again with current technology, the telephone industry undoubtedly would do many things differently. Wiring to subscribers might be in the form of a time division multiplex loop from house to house, rather than a separate pair from each residence to the central office. The bandwidth would certainly be wider; signaling and ringing would be different. But for most systems, the luxury of an entirely new design is not possible. There is not enough capital, manufacturing capacity, or manpower available in the United States to completely rebuild the telephone system, nor is there any reason to want to do so.

The improvements described in this and following chapters will be implemented in the way the telephone companies have managed change for over a hundred years; that is, only small changes which are compatible with the existing network at the interface (the point where it is connected) are made at any given time. Change in the telephone industry is evolutionary, not revolutionary; subtle, rather than sudden.

DC REQUIREMENTS FOR THE LOCAL LOOP

In order to discuss the changes to the circuits in the telephone set, we must consider the portion of the system that we cannot change, but with which the electronic telephone set must interface. This is, of course, the local loop from the central office. It is important that, where possible, the electronic telephone use the loop current from the telephone company central office to provide power for the electronic circuits. Therefore, let's first examine the dc parameters of the local loops seen by the telephone set.

In *Figure 3-1* are shown typical battery feed circuit arrangements used in central offices to connect to the local loop. *Figure 3-1a* shows a capacitive coupled circuit and *Figure 3-1b* a transformer coupled circuit. The configuration at the central office depends on the type of supervisory signaling used on the transmission lines being connected and on how the central office was designed.

In each battery feed arrangement, L1 and L2 are equal and their sum is the total inductance of the feed circuit in the central office including the sensing relay. The same holds true for the resistance; R1 and R2 are equal and their sum is the total resistance, again including the sensing circuit being used. R_L and C_L are the lumped values for the line resistance and line capacitance. R_{eq} is the equivalent resistance that the telephone set presents to the line for determining the local loop current. V_B is the central office voltage applied to the local loop. *Table 3-1* lists some typical values for the battery feed components in several different systems.

For most conventional telephone sets in the United States, the component which determines the minimum operating current in the local loop is the carbon transmitter. It requires 23 mA to operate reliably, but minimum current on the local loop is usually set at 20 mA. (The sensitivity of the line relays in the central office also can determine the minimum local loop current; however, they are designed to operate within the 20 mA limit of *Table 3-1*.) With V_B at 48 volts and R_T = 400 ohms, the maximum current that would flow in the local loop is 120 mA if the line were shorted at the central office. This is system 1 of *Table 3-1*. (R_T varies considerably with different countries; e.g., France has 380 to 780 ohms.)

The equivalent resistance, R_{eq}, of *Figure 3-1* for the telephone set is typically about 400 ohms (but may be as little as 100 ohms). Using Ohm's law, if 48 volts is divided by 23 mA (the minimum current to operate the carbon transmitter), the maximum local loop resistance is 2,087 ohms. If R_{eq} = 400 ohms and R_T = 400 ohms, the maximum line resistance, R_L, for the local loop is 2,087 − 800 = 1,287 ohms, or typically 1300 ohms (see *Table 1-6*).

If electronic telephone sets are used with transmitters that are microphones other than the carbon granule type, then the minimum loop current may be much less than 20 mA, say 10 mA. In this case, the limiting component specification is the minimum current required to operate the line relay in the central office.

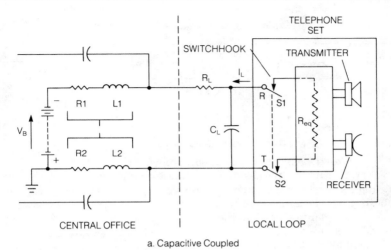

a. Capacitive Coupled

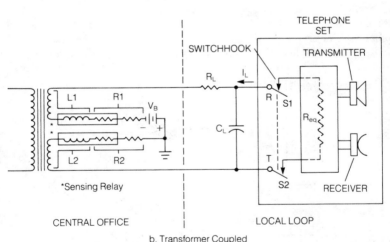

b. Transformer Coupled

Figure 3-1. *Typical Central Office Battery Feed*

Under these conditions, the on-hook current drawn by the electronics inside the telephone set must be well below the minimum current required by the line relay, otherwise the line relay would energize and incorrectly indicate an off-hook condition. Obviously, the on-hook current drain also must be held to a minimum so the central office supply will not be overloaded. For these reasons, some local exchanges have specifications for minimum loop current that fall in the range from 6 mA to 25 mA. This is especially true for local exchanges that may be detecting off-hook loop current with optoelectronic couplers, which are then interfaced directly to digital logic circuits.

Table 3-1. *Central Office Battery Feed*

System	Nominal Battery Voltage V_B (volts)	Typical Resistance (Ohms) $R_T = R_1 + R_2$	Current (mA)	
			I_{MAX}	I_{MIN}
1	48 ± 2	400	120	20
2	48 ± 2	800	60	20
3	24 ± 2	400	60	20

TWO-WAY SPEECH BLOCK DIAGRAM

With the past discussion as background, let's now shift to the main topic of this chapter—the electronic circuits and different devices that replace the transmitter, the receiver and the transformer coupled induction coil (or hybrid) of the conventional telephone set.

In order to relate how the electronics fit into the telephone set, the equivalent circuit of *Figure 2-15* for the hybrid circuit is shown again in equivalent form in *Figure 3-2*, modified with the two-way speech electronics. The transmitter now feeds signals to a transmission amplifier marked with the symbol A_T, and the receiver is driven by a receiver amplifier marked with the symbol A_R. The triangle is the common symbol for an electronic amplifier. The amplifiers now couple to the equivalent circuit at the same point in *Figure 3-2* as the transmitter and receiver did in *Figure 2-15*. The balance network of *Figure 2-15* is represented by Z_B in *Figure 3-2*.

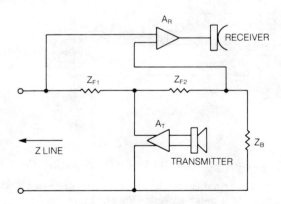

Figure 3-2. *Equivalent Circuit of an Amplified Speech Circuit*

Two more circuits that are not shown in *Figure 3-2* are required when electronic circuits are used in the telephone. These additional circuits, the overvoltage protection and polarity protection circuits, are shown in the block diagram of *Figure 3-3*.

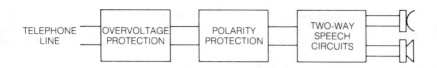

Figure 3-3. *Block Diagram of Speech Section of Telephone Set*

PROTECTION CIRCUITS

Overvoltage Protection

Electronic circuits using small signal transistors and integrated circuits are easily damaged by high voltage transients that may appear on the line due to lightning, switching transients, or power line induction. Therefore, some form of transient and overvoltage protection must be used to protect the circuits in the electronic telephone set. Since the technology used for making integrated circuits cannot provide high enough breakdown voltage, one or more external zener diodes are placed across the line to provide overvoltage protection. When the input voltage exceeds the zener (breakdown) voltage, the zener diode conducts and holds the voltage input to the electronic circuit at the rated zener voltage. If the overvoltage protection is needed at a point where either polarity voltage may appear, the device may be constructed as two zener diodes connected back-to-back, or two physically separate zener diodes may be connected back-to-back across the line.

Polarity Protection

Polarity of the normal input voltage is critical for electronic circuits since they won't operate if the polarity is reversed, and they may be damaged. The method commonly used to protect against polarity reversal is the rectifier bridge. The output voltage polarity of a rectifier bridge is always the same regardless of input voltage polarity.

Conventional Rectifier Bridge

A rectifier bridge can be built easily by using four diodes as shown in *Figure 3-4a*. Since the rectifier diodes can withstand high voltage transients, only one zener diode is required on the bridge output to protect the speech circuits.

A conventional full-wave silicon diode rectifier bridge has a forward voltage drop of 1.5 volts. This voltage drop becomes important for electronic telephone sets because they are voltage-operated rather than current-operated as the conventional telephone set discussed in Chapter 2. Therefore, the minimum line voltage is a much more important specification for the electronic telephone set. Since the speech circuits must have 3.5 volts to operate, the line voltage must be at least 5.0 volts if a silicon diode bridge is used.

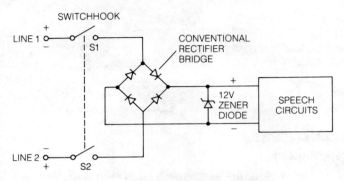

a. For Line Voltage of 5 Volts or Higher

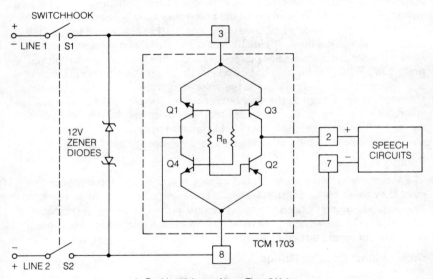

b. For Line Voltage of Less Than 5 Volts

Figure 3-4. *Over voltage and Polarity Protection Circuits*

Low-Voltage Rectifier Bridge

Some telephone systems may have less than 5.0 volts available. In these cases, the electronic telephone set must use a low-voltage rectifier bridge which has a voltage drop of 0.5 volt or less. This characteristic can be obtained with an active bridge arranged as shown in *Figure 3-4b* so that the voltage drop between input and output is the $V_{CE(sat)}$ of a transistor. $V_{CE(sat)}$ is the voltage drop across the collector-emitter junction when the junction is in a condition where it is switched ON. This operating condition is called saturation. Depending upon the polarity of the input voltage applied to pins 3 and 8, either transistor pair Q1-Q2 or Q3-Q4 conduct. If pin 3 is positive and pin 8 is negative, Q3 and Q4 conduct. Q1 and Q2 conduct when pin 8 is positive and pin 3 is negative. Thus, the output voltage polarity at pins 2 (positive) and 7 (negative) is the same regardless of the line polarity. The values of resistors R_B are chosen so that the transistors always operate in saturation. The voltage drop across the bridge is the sum of the $V_{CE(sat)}$ of the two "ON" transistors. It is about 0.4 volt in the Texas Instruments TCM1703 integrated circuit low-voltage rectifier bridge.

This bridge circuit also has the desirable characteristics of low shunt current (about one milliampere) and no effect on speech frequencies. However, the maximum line voltage it can handle is about 14 volts. This is sufficient for the normal voltage variations, but high-voltage transients would damage it. Therefore, the transient protection circuit must be placed between the bridge and the line input as shown in *Figure 3-4b*. Since the input voltage may be of either polarity, two zener diodes are connected back-to-back across the line.

THE SPEECH CIRCUIT

A simplified diagram of a speech circuit suitable for implementation in integrated circuit form is shown in *Figure 3-5a*. It is connected to the telephone line via the low-voltage rectifier bridge as shown in *Figure 3-5b*. The bridge is equivalent dynamically to a small resistance in series with the signal path and a high resistance in parallel with it. The interconnection of *Figure 3-5b* is arranged just like the hybrid decoupling equivalent circuit of *Figure 3-2*. The external components that must be added are identified as well as the external connections.

In *Figure 3-2*, the receiver amplifier A_R is decoupled from the transmitter amplifier A_T by the hybrid bridge consisting of impedances Z_{F1}, Z_{F2}, Z_B, and the line impedance Z_L. Therefore, just as in the conventional telephone set, the receiver amplifier receives signals from the telephone line, but not from the transmitter amplifier. In like fashion, the transmitter amplifier sends a signal to the line and the only signal present at the receiver is the sidetone.

Note that the amplifiers in *Figure 3-5* are shown to have variable gain; thus, the gain of the amplifiers can be adjusted according to the length of the subscriber loop to compensate for line loss. The sidetone level can be varied by adjustment of the external balancing components R6, R7 and C5. The transmitter amplifier A_T can be designed to accommodate either high impedance (electret) or low impedance (electrodynamic) microphones. Resistors R1 and R2 are hybrid bridge resistors that replace induction coil windings in the conventional telephone set. Capacitor C1 corrects the frequency response.

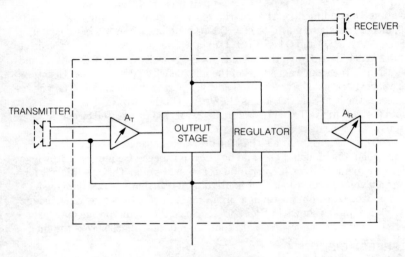

a. Integrated Circuit

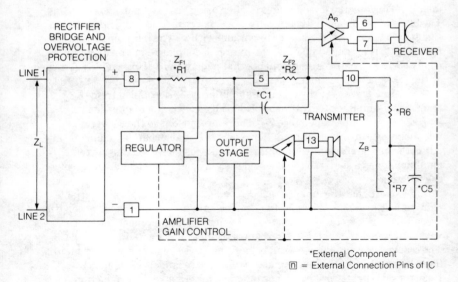

*External Component

$\boxed{n}$ = External Connection Pins of IC

b. Interconnected to Line

Figure 3-5. *Two-Way Speech Integrated Circuit*

Regulator

The regulator controls the dc voltage/current characteristics of the entire two-way speech integrated circuit. The stabilized output voltage is supplied to the amplifiers A_T and A_R and to other circuits requiring it throughout the integrated circuit. The regulator also supplies the gain control signal to amplifiers A_T and A_R, as well as control signals (biasing) to a number of the current generators. The regulator block diagram is shown in *Figure 3-6*. The functional parts of the regulator are described below.

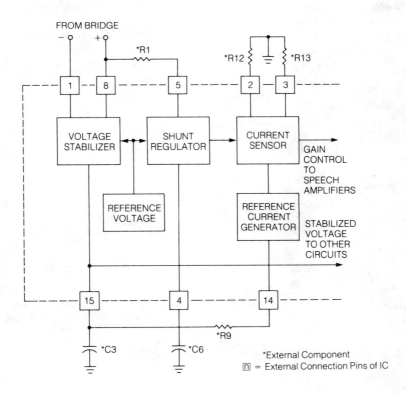

Figure 3-6. *Regulator Block Diagram*

Reference Voltage (Band Gap)

The reference voltage (band gap) is a dc voltage which is relatively constant over a wide range of variations in applied dc voltage and ambient temperature. It is the reference used by the voltage stabilizer and shunt regulator to provide a stable voltage over varying current conditions.

Voltage Stabilizer

The Voltage Stabilizer, as shown in *Figure 3-7*, is a common series regulator. An operational amplifier (op amp) compares a sample of the output voltage with the reference voltage. The output of the op amp is dependent on the error signal input which is the difference between the reference voltage and the sampled output voltage. It varies the base bias for transistor Q1; thereby varying the voltage drop across Q1 from emitter to collector. The circuit operates with Q_1 in the linear range. That means that the voltage from emitter to collector must be above $V_{CE(sat)}$ of Q_1. The regulator cannot eliminate all of the variation. The remaining voltage changes at the output are called ripple. The ripple in the output is smoothed by an external capacitor connected to pin 15.

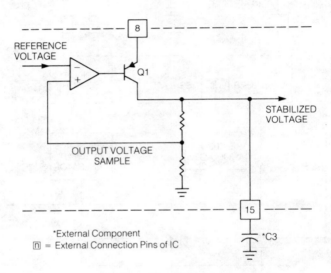

Figure 3-7. Voltage Stabilizer

Shunt Regulator

As the name suggests, the Shunt Regulator shown in *Figure 3-8* varies the dc current that is shunted through a parallel path as the current drawn by the total integrated circuit varies so that the total current load on the regulator remains fairly constant. An operational amplifier compares the reference voltage to a sample of the output voltage, as is done in the voltage stabilizer. The op amp output, which is proportional to the difference between the reference voltage and the output voltage sample, drives the base of Q1 and controls the current that is shunted through Q1.

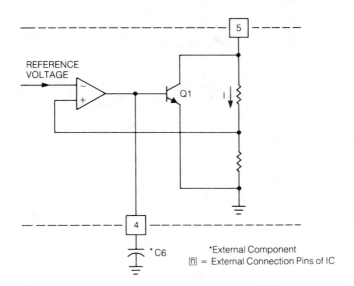

Figure 3-8. Shunt Regulator

Note in *Figure 3-5* that pin 5 is connected to the telephone line through resistor R1. In order to produce a dc characteristic which meets the specifications of the various telephone companies, the regulator must have a rather low internal impedance of around 50 ohms for dc and low frequencies. However, this low impedance must not occur at frequencies within the voice band; if it did, the speech power would not leave the telephone set. The circuitry of the shunt regulator is designed so that the response of the regulator is well below 300 Hz; thus, the regulator will not shunt speech frequencies above 300 Hz.

Reference Current Generator

The Reference Current Generator, shown in *Figure 3-9*, uses current mirror transistors Q3 and Q4 to ensure that transistors Q1 and Q2 carry the same current. The value of the reference current I_{REF} is independent of the supply voltage, but is determined by R9, an external precision resistor.

The circuit of *Figure 3-9* depends on the matching of transistor characteristics. This is a significant advantage in integrated circuits where the transistors are made by modifying silicon material by diffusing impurities into the silicon. Since they are small, very close to each other and made at the same time, they come out with almost identical characteristics.

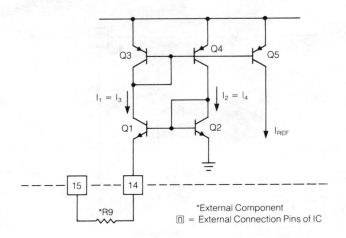

Figure 3-9. *Reference Current Generator*

As a result, the I_{REF} that flows in the collector of Q5 is the same as the current I_4 that flows in the collector of Q4 because the V_{BE} and gain characteristics of Q4 and Q5 are the same. The resistor R9 sets I_1, the collector current of Q1. Due to the circuit connection of the NPN transistors Q1 and Q2, the collector current I_2 is equal to the collector current I_1. Since the circuit formed by PNP transistors Q3 and Q4 is a mirror image of the NPN circuit (PNP transistors operate with reverse polarity from NPN), the collector current I_4 is equal to the collector current I_3. The interconnection of the circuit forces $I_1 = I_3$ and $I_2 = I_4$; therefore, the complete circuit results in a very stable I_{REF} set by the resistor R9.

Line Current Sensor and Comparator

The reference current feeds into the line current sensor circuit shown in *Figure 3-10*. The output controls the speech amplifier gain and varies the gain according to the length (resistance) of the local loop feeding the telephone set. As discussed previously, line current decreases as the length of the local loop increases. In other words, it varies inversely with the length of the telephone line. It, therefore, can be used as a measure of the line length. By comparing the loop current with the reference current, a control signal is developed that increases the speech amplifier gain for longer lines. The two external resistors, R12 and R13, determine the threshold at which gain control starts. A special case exists if pin 2 is connected directly to ground (R12 = zero). In this case, the speech circuit gain is independent of line length and is not regulated.

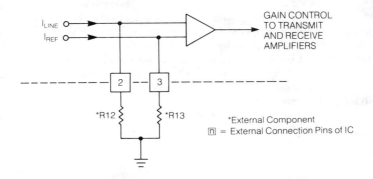

Figure 3-10. *Line Current Sensor*

Transmitter Amplifier

Figure 3-11 shows the block diagram of the transmitter amplifier, A_T. It matches the microphone to the telephone line by providing gain and correct output impedance. Two versions of the integrated circuit are required; one for the high impedance electret microphone and another for the low impedance electrodynamic microphone.

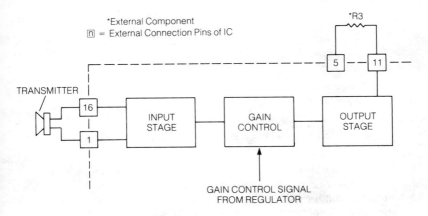

Figure 3-11. *Transmitter Amplifier A_T*

High Impedance Input Stage

The input stage shown in *Figure 3-12* is the one used with the electret microphone. The input signal is coupled to Q1, a field-effect transistor called a junction FET. The three connections are called gate, drain and source. The field-effect transistor uses electric fields between the gate and source to control current between drain and source. Because it is a field-controlled device, it has a high input impedance to signals coupled to the gate input. The symbol in the drain circuit of the FET indicates a constant current source. The circuitry of a constant current source is very similar to that shown in *Figure 3-9* for the reference current generator. This symbol is used instead of all the components in order to simplify the figure.

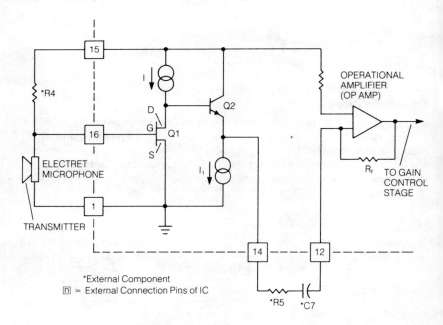

Figure 3-12. *Input Stage for High Impedance Microphone*

The amplified signal from the FET is coupled to Q2, an emitter follower which has a constant current source in its emitter. The emitter follower has no voltage gain, but provides current gain as it couples the signal to an operational amplifier represented by the triangular symbol. Many components are required to make up an operational amplifier (a typical one is shown in *Figure 3-13*); therefore, the symbol is used in order to simplify the figure.

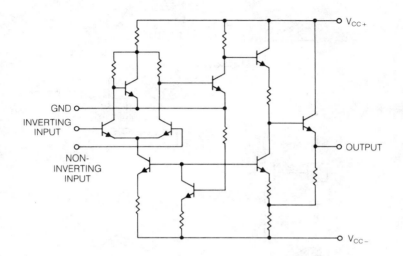

Figure 3-13. *Typical Operational Amplifier*

The gain of the input stage is controlled by varying the value of the external resistor R4. The operational amplifier gain is determined by the ratio of Rf to R5. Since R5 is external, it may be changed to adjust the gain. C7 is used for blocking dc.

Low Impedance Input Stage

If the microphone is electrodynamic, then the input stage changes to that shown in *Figure 3-14*. This circuit uses two operational amplifiers connected in cascade. The gain is controlled by the value of external resistor R4, whose value is also the input resistance as seen from the microphone. Each operational amplifier again is represented by a triangular symbol, but actually is made up of a number of components similar to those shown in *Figure 3-13*.

Gain Control Stage

The input stage couples the signal to the gain control stage. It again is an operational amplifier that has its basic gain set by the ratio of R_f and R_{IN}, but whose gain is controlled by the signal from the regulator that is output by the line current sensors. The control signal is proportional to the line length; thus, the gain controlled amplifier's gain is automatically adjusted to compensate for line length to overcome line loss.

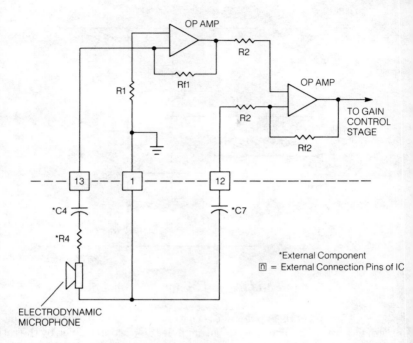

Figure 3-14. Input Stage for Low Impedance Microphone

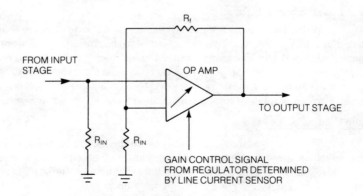

Figure 3-15. Gain Control Stage

Output Stage

The Output Stage is shown in *Figure 3-16*. It is a current generator driven by the gain control stage. External resistor R3 not only fixes the quiescent current of the output stage, but also determines the ac impedance of the speech circuit as seen from the telephone line.

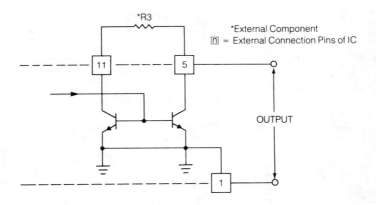

Figure 3-16. *Output Stage of A_T*

Equivalent Circuit Analysis

To understand how the output stage couples to the line, refer to the diagram of *Figure 3-17*, which is *Figure 3-5* with the regulator and the receiver amplifier removed. These two circuits have a high impedance; therefore, they do not load the circuit and do not affect the analysis here. The output stage is represented in *Figure 3-17* as a current generator in parallel with an internal impedance of Z_o.

The impedances Z_L, R1, R2 and Z_B form a bridge. This bridge is balanced when:

$$Z_L \times R2 = Z_B \times R1$$

The balance condition ensures that the microphone and receiver are decoupled.

The value of the ac impedance seen looking into the circuit from the line is:

$$Z_{IN} = R1 + \frac{Z_o(R2 + Z_B)}{Z_o \times (R2 + Z_B)}$$

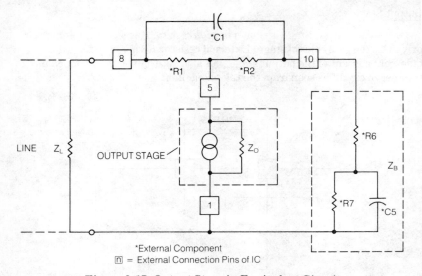

Figure 3-17. *Output Stage in Equivalent Circuit*

This impedance, Z_{IN}, must be variable by the manufacturer or installer so that the equipment may be used with networks with different requirements. But note that both balance and Z_{IN} depend upon the value of Z_B. This is not desirable. Z_{IN} can be made essentially independent of Z_B if Z_B is some multiple K of Z_L, and R2 is the same multiple of R1.

For example, let

$$Z_B = KZ_L \text{ and } R2 = KR1$$

If the value of K is chosen sufficiently large such that KZ_L is much greater than Z_o, then Z_{IN} is given by:

$$Z_{IN} = R1 + \frac{Z_o(KR1 + KZ_L)}{Z_o + KR1 + KZ_L}$$

Now, since KZ_L is much greater than Z_o, Z_o in the denominator has little effect on the result; therefore, it can be omitted. With that done, the $(KR1 + KZ_L)$ factors cancel, as follows:

$$Z_{IN} = R1 + \frac{Z_o(KR1 + KZ_L)}{(KR1 + KZ_L)}$$

then

$$Z_{IN} = R1 + Z_o$$

Using the above criteria, the ac impedance seen from the line depends mainly on the output impedance of the output stage which may be easily adjusted by the value of external resistor R3 in *Figure 3-16*.

Receiver Amplifier

The receiver amplifier block diagram with external components is shown in *Figure 3-18*. It consists of an Input Stage, Gain Control Stage and Output Stage. Let's discuss each of these separately.

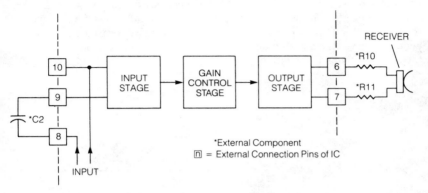

Figure 3-18. *Receiver Amplifier, A_R, Block Diagram*

Input Stage

The input stage is shown in *Figure 3-19*. The input signal comes from the end of elements R1 and R2 of the hybrid bridge (*Figure 3-5*) and is fed through pin 8 and capacitor C2 to the receiver amplifier. The connection arrangement of the two amplifiers shown in *Figure 3-19* is for dc biasing purposes only. The amplifiers are op amps with their gain set by feedback through the resistor ratio of R_f and R_{IN}. As before, the amplifier symbol represents a significant number of components.

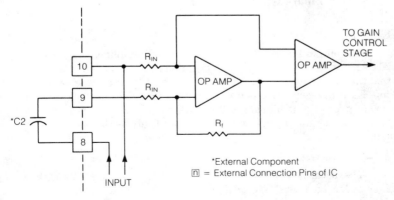

Figure 3-19. *Receiver Amplifier Input Stage*

Gain Control Stage

The gain control stage shown in *Figure 3-20* is essentially the same as for the transmitter amplifier. It is an operational amplifier that has its gain controlled by the signal from the line current sensing circuit in the regulator so that the gain is automatically adjusted to compensate for the length of the local loop; thus, for the line loss. However, there is one significant difference from the transmitter amplifier; the output is double-ended rather than single-ended. It drives the output stage in push-pull (one output goes positive while the other output goes negative).

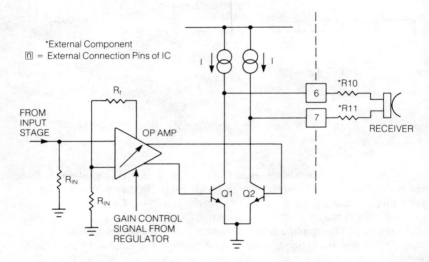

***Figure 3-20.** Gain Control and Output Stages*

Output Stage

The output stage in *Figure 3-20* is a much simplified version of the actual circuit in the two-way speech integrated circuit. The important points are that the output stage is a push-pull stage and drives the receiver in that fashion. External resistors are in series with the receiver to allow adjustment of gain and overall dynamic range. The stage must be designed to have a low output impedance and the output drive is limited only by the maximum available current.

LINE BALANCING

The discussion of the telephone set would not be complete without considering line balancing. We shall discuss this topic from a general view, considering both electrical and acoustic factors.

The entire telephone set, including the handset, the amplifiers, and the hybrid, is shown in *Figure 3-21*. A signal a from the microphone results in the a–b portion being delivered to the telephone line amplified by a factor K based on the gain of A_T and the ratio of decoupling of the hybrid. The imbalance of the hybrid results in a fraction of this signal, b, being fed through the receiver amplifier, A_R, to the receiver. This is the sidetone as discussed in Chapter 2.

The electrical signal to the receiver is converted into sound and fed to the user's ear. However, a fraction, b_x, of that acoustical signal will be fed back to the microphone via the air and possibly the facial bones of the user. This acoustical feedback can be positive and add to the original signal, a, to cause howling if the level of feedback is high enough. The acoustical feedback can be reduced by the shape of the handset, the mounting of the receiver and microphone inside it, and the material from which it is made.

The electrical signal balancing depends on the hybrid network and the value of line balancing impedance Z_B. The latter values are determined by the telephone company after a great deal of statistical survey on the lines used by the telephone system and on the spread of parameters of the production components that go into the telephone set.

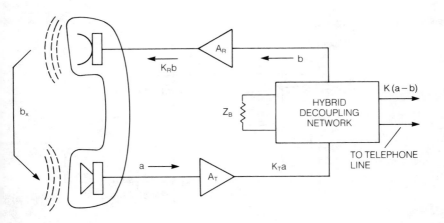

Figure 3-21. *Signal Paths*

TELEPHONE SET VOLUME COMPENSATION

Recall in the discussion of the conventional telephone set that there were varistors (components whose resistance changes as the current through them changes) in the hybrid circuit that adjusted their resistance so that the volume of the sound in the telephone system from transmitter to receiver stayed about the same even though the length of the local loop increased. *Figure 3-22* illustrates this feature for the common Bell Type 500 telephone set. Note that the horizontal axis is either local loop resistance R_L in ohms (top of chart) or line length in thousands of feet (bottom of chart). The vertical axis is the response in decibels.

The curve of the relative response of the Bell Type 500 telephone set receiver shows that the relative volume level delivered to the receiver stays within ± 1 dB for a local loop length of 0 feet ($R_L = 0$) to a length of 14,000 feet ($R_L = 1180$ ohms). As a comparison, an older telephone set, the Type 302, had a 5 dB loss in volume when the local loop was 12,000 feet long. Similarly, another curve in *Figure 3-22* shows the transmitting response. As explained for the two-way speech integrated circuit, a signal from the regulator based on the current in the local loop adjusts the gain of the transmitter amplifier and the receiver amplifier of the electronic telephone to provide the same or better compensation than that of the Bell Type 500 set.

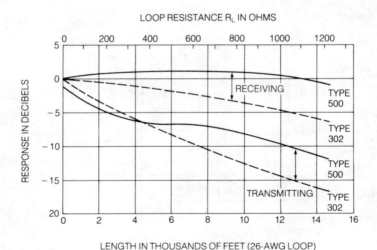

Figure 3-22. *Conventional Telephone Set Relative Volume Levels vs Line Length*
(Courtesy of Bell Laboratories)

3

APPLICATION OF SPECIFIC TWO-WAY SPEECH CIRCUITS

Various manufacturers market integrated circuits which perform the two-way speech circuit function. Two such circuits are the Texas Instruments TCM1705 and TCM1706. The 1705 is designed for a low impedance electrodynamic microphone and the 1706 for a high impedance electret microphone. Both circuits use a low impedance electrodynamic receiver.

The applications circuit diagrams for these two ICs are shown in *Figures 3-23* and *3-24*. Full details of the performance of the circuits are available on published data sheets. A summary of the primary characteristics is as follows:

Table 3-2. *Primary Characteristics of IC*

	Minimum	Maximum	Conditions
Line Voltage	3.5 volts		
Transmitter Input Impedance	200 ohms	2,000 ohms	
Transmitter Amplifier Voltage Gain		48 dB	Battery = 48V $R_L = 800\Omega$, Mic = 350Ω, Rec = 350Ω, f = 1 kHz, $V_T = 3mV$ $R_{exchange} = 800\Omega$
Distortion (at speech frequencies)		3%	f = 1 kHz, I_L = 18 to 80mA $V_L = 1.3V$ peak $V_R = 400mV$ peak
Receiving Voltage Gain		− 11.4 dB	Battery = 48V $R_L = 800\Omega$, Mic = 350Ω, Rec = 350Ω f = 1 kHz $V_L = 0.3V$, $R_{exchange} = 800\Omega$
Receiving Power Gain		7.5 dB	
Line Impedance Limits	550 ohms	850 ohms	

Note that the receiving voltage gain is given in minus decibels. This means that the voltage which appears across the receiver is less than the voltage received across the line. However, there is a positive power gain so that more power is delivered to the receiver than the power of the signal from the line at the input to the receiver amplifier.

To emphasize the flexibility of the two-way speech integrated circuit, *Table 3-3* gives the adjustments that can be made with the various components shown in *Figure 3-22* and *3-23*.

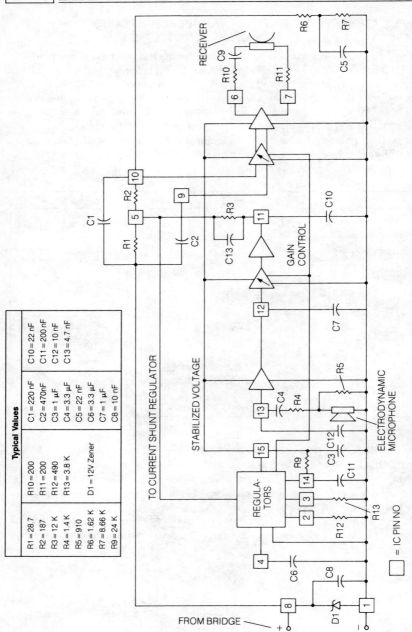

Figure 3-23. Application Circuit Diagram for TCM1705

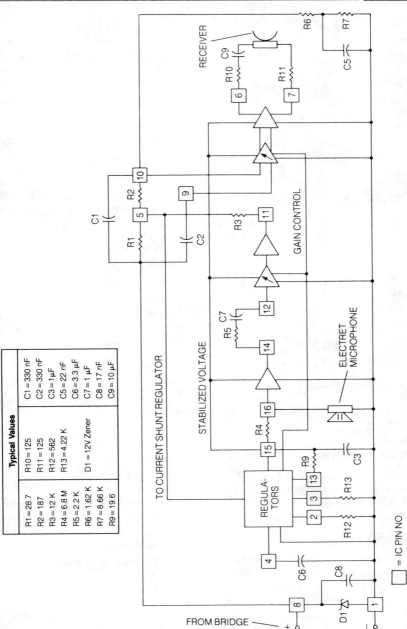

Typical Values

R1 = 28.7	R10 = 125	C1 = 330 nF
R2 = 187	R11 = 125	C2 = 330 nF
R3 = 12 K	R12 = 562	C3 = 1 μF
R4 = 6.8 M	R13 = 4.22 K	C5 = 22 nF
R5 = 2.2 K		C6 = 3.3 μF
R6 = 1.62 K	D1 = 12V Zener	C7 = 1 μF
R7 = 8.66 K		C8 = 17 nF
R9 = 19.6		C9 = 10 μF

Figure 3-24. *Application Circuit Diagram for TCM1706*

Table 3-3. *External Component Function*

Component	Function
R1, R2	Set the receiver gain which is proportional to their values.
R6, R7, C5	Determine the value of the balancing network.
R4, R5	Adjust both the transmitting gain and the matching impedance for the microphone.
R9	Fixes the bias current for the whole circuit. With R9 = 24K, the minimum operating current for the circuit is 12mA. By decreasing R9, the maximum operating current increases and a better receiving and transmitting dynamic can be achieved.
R10, R11	Adjust both the receiving gain and the matching impedance for the headphone.
R12, R13	Adjust the line length regulation function according to the used feeding system, after R9 has been fixed.
R3	Fixes the resistive part of the line impedance and matches the output stage to the line.
C1	Provides a flat receiving frequency response.
C2	DC blocking capacitor on the receiving amplifier input.
C3	Filter capacitor of the internal series voltage regulator.
C4	DC blocking capacitor on the transmitting amplifier input.
C6	Filter capacitor of the internal shunt regulator.
C7	Decouples the second stage of the transmitting amplifier.
C8	Fixes the capacitive part of the line impedance and provides good radio frequency rejection.
C10, C11, C12, C13	Provides radio frequency rejection.
D1	Overvoltage protection zener.
C9	Optional dc blocking capacitor can be added in series with the receiver in the receiver output stage to increase the output voltage range by 100mV peak.

WHAT HAVE WE LEARNED?

1. Changes to the telephone set must be made in a manner that maintains the proper characteristics for the line interface.
2. The coupling of two–way speech circuits into the telephone set is similar to the conventional hybrid circuit except a resistive bridge replaces the induction coil.
3. Two-way speech circuits are made with integrated circuits because they offer high performance, low power consumption, small size, are lightweight and have high reliability.
4. Two-way speech circuits must operate from very low voltage (below 4V) if powered from the telephone line.
5. Overvoltage protection must be provided to prevent damage to electronic circuits by voltage variations and spikes on the line.

6. The two-way speech circuits must operate with either polarity voltage on the telephone line. A rectifier bridge is used to provide a constant polarity voltage.

7. Two-way speech circuits must draw very low leakage current from the telephone line in the on-hook condition.

8. Circuits for voltage and current regulation are contained in the two-way speech integrated circuit.

9. Controlled gain is provided by the two-way speech amplifiers in the electronic telephone set to compensate for the line length. This is in contrast to the conventional telephone set which compensates for line length by using varistors.

10. Either an electrodynamic or an electret microphone may be used with the proper choice of amplifier input.

Quiz for Chapter 3

1. The equipment used in the telephone network changes:
a. rapidly, to keep up with new technology.
b. almost never, because it is doing the job.
c. slowly, to balance new technology against cost.

2. The telephone set can change:
a. very little, because it is connected to a network that changes slowly.
b. as long as the interface to the subscriber loop remains the same.
c. only in cosmetics, such as color, shape, style, etc.

3. The carbon transmitter:
a. is a comparatively recent electronic development.
b. was invented by Thomas A. Watson, Bell's assistant.
c. determines the minimum loop current.

4. On-hook current must be kept low so that the:
a. line relays in the CO will not mistake it for off-hook current.
b. comparatively small wires in the cable will not overheat.
c. ringer will not ring incorrectly.
d. carbon microphone will not be damaged.

5. Power to operate the telephone and the subscriber loop is usually provided by:
a. the telephone company, and is low-voltage ac.
b. the subscriber, through the use of batteries in the set.
c. the CO, and is called battery feed because batteries are sometimes used.

6. Overvoltage protection is:
a. needed in the telephone set because of the high ringing voltage.
b. is nearly always incorporated on the IC chip.
c. is primarily a central office function.
d. needed because of transients from dial pulsing, lightning, induction, or short circuits.

7. A special rectifier bridge is used in electronic telephones because:
a. the ac power must be converted to dc to simulate a battery.
b. it is necessary to shunt the speech frequencies and keep them off of the loop.
c. the voltage drop across conventional bridges leaves too little voltage to operate the set.

8. The band gap:
a. regulates the difference between the high and low DTMF signaling tones.
b. provides a dc voltage reference for the stabilizer and regulator.
c. regulates the ringing voltage on the subscriber loop.
d. is the power source for the IC's in the electronic telephone.

9. The transmitter amplifier:
a. matches the microphone impedance and gain to the line.
b. provides a built-in loudspeaker in the telephone set.
c. is part of the network, not part of the telephone set.

10. The receiver amplifier:
a. provides a built-in loudspeaker in the telephone set.
b. automatically compensates for the length of the subscriber loop.
c. is provided for use of hearing-impaired customers.
d. is powered by an internal battery in the telephone set.

Electronic Dialing and Ringing Circuits

ABOUT THIS CHAPTER

In this chapter, the discussion centers on replacing the conventional telephone set components that perform the functions of dialing and ringing with electronic circuits that do these functions. One might say that electronics has made the mechanical rotary dial obsolete. Solid-state integrated circuits are available that generate the standard dial pulse sequence when a telephone number is entered by depressing keys on a keypad. Electronic memories can retain one or more phone numbers so that the touch of only one key on a keypad sends the entire number down the line. Likewise, electronic ringers are replacing the bell to signal an incoming call. Single-tone and multitone electronic ringers are available.

DIAL PULSE GENERATION USING INTEGRATED CIRCUITS

Dial pulsing is the most commonly used method in the world for sending addresses (telephone numbers) to the central office. Consequently, the mechanical rotary dial with its gears, cams, contacts, and rotary speed governor was one of the first targets for the substitution of electronics. The rotary dial itself is being replaced with a pushbutton keypad and the pulse generating mechanism is being replaced with an integrated circuit called an impulse dialer. Several versions of this circuit are made by a number of semiconductor manufacturers.

In order to be compatible with electromechanical central offices, the electronic impulse dialer must send dial pulses at the 10 pulse per second rate. Since a person can push the buttons faster than the pulses can be sent, an electronic memory capable of holding from 17 to 20 digits is included in the circuit to hold the number while the dial pulses are transmitted. The number remains in this memory for several hours or until another number is entered.

Redial

With the memory function available, it was relatively easy to incorporate a redial function. By pushing only one button, the last number entered will be redialed. This feature is helpful when continually trying to reach a busy number. Initially, the redial feature met with some resistance from the telephone companies because of fears of clogging the network because redialing could be done so rapidly and frequently. Also, they were concerned about the maintenance of batteries used to power the electronic circuits in the telephone set. (Remember the station battery at each telephone was abandoned in favor of a common battery at the central office exchange very early in the 1900's because of cost and maintenance.) The development of complementary metal-oxide-silicon (CMOS) semiconductor technology, which produces integrated circuits with very low power consumption, has solved the latter problem. The current drain of these circuits is only a few microamperes so that it is possible to operate the electronics from the dc power available from the telephone line. However, if a battery of the type commonly used to power calculators or other consumer electronic goods were used to provide only the very small current required by the number memory, the battery would last for several years.

CIRCUIT POWER AND VOLTAGE TRANSIENTS

Before proceeding with the description of the electronic pulse dialer, there are two general subjects that need to be covered in a bit more detail. One is circuit power and the other is voltage transients.

Circuit Power

Figure 4-1 shows a number of ways that power is provided to electronic circuits inside the telephone set that are powered from the line. The ringer circuit must be supplied from the line ahead of the switchhook so that the circuit can be energized by the ringer signal even though the telephone handset is on-hook. In this case, a diode rectifier bridge must precede the circuits (if it is not provided in the ringer circuit) to protect against line voltage polarity reversal.

The dial circuits have their power supplied through the switchhook. Again, a diode rectifier bridge must be supplied externally if it is not provided in the integrated circuits.

Still other circuits, such as the speech circuits discussed in Chapter 3, have their power supplied through the low-voltage rectifier bridge (TCM1703) which handles the polarity problem. Instead of other circuits having independent diode bridges, they can be powered from the low-voltage rectifier. This is shown by the dotted lines to the dial circuits.

Also, it must be remembered that some electronic telephones will have only certain of the functions replaced by electronics and other parts will be the same as conventional telephone sets. For this reason, low-voltage rectifier bridges or standard diode bridges may have to be added in the power supply leads that feed the electronic circuits.

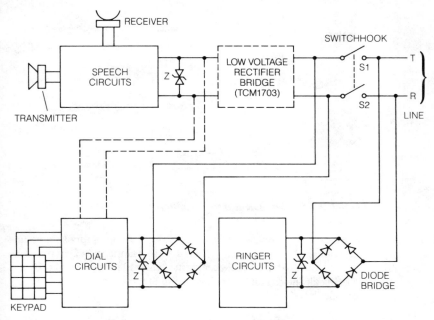

Figure 4-1. *Circuit Power for Line Driven Circuits*

Voltage Transients

When the handset is on-hook, the switchhook is open and the full exchange battery voltage (usually 48 Vdc) appears at the tip and ring terminals of the telephone set. When the handset is lifted, the switchhook closes. If there is any contact bounce (momentary closures and openings) and the circuit in the telephone set is inductive, relatively high voltage pulses called transients will be generated. Dial pulsing also produces these high voltage transients.

Figure 4-2 shows an equivalent circuit of such a case just as the switch in the circuit has opened. A current I had been flowing. Due to the magnetic field stored in the inductance L, which collapses when the current stops, a voltage is generated across the inductance that appears across the opening switch contacts. This voltage can be calculated by the following equation:

$$V = \frac{L \, \Delta I}{\Delta t}$$

where

   L = inductance (in henries) in circuit
   ΔI = the change in current (in amperes) that occurred
   Δt = the time (in seconds) it took to make the current change

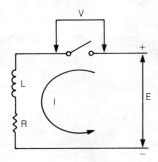

Figure 4-2. *Equivalent Circuit for Producing Voltage Transients*

An example will show how large the voltage can be. Suppose L equals 1 henry of inductance, that 10 milliamperes (10×10^{-3} amperes) of current were flowing when the switch opened, and that it took 10 microseconds (10×10^{-6} seconds) to break the circuit. By substituting in the above equation, the voltage can be calculated.

$$V = \frac{1 \times 10 \times 10^{-3}}{10 \times 10^{-6}}$$
$$= 1 \times 10^{3}$$
$$= 1,000 \text{ volts}$$

The voltage generated is 1,000 volts. Integrated circuits and many small transistors cannot withstand this much voltage and will be permanently damaged. Thus, the electronic phone must have high-voltage transistors in circuits that interface to the line and must have circuits to suppress or absorb these transient voltages. The overvoltage protection, speech-muting, and ringer-muting (anti-tinkle) circuits perform this function.

Overvoltage protection was discussed in Chapter 3. Recall that either single-polarity or double-polarity zener diodes are used depending on the expected polarity of the voltage at the point of protection. The speech-muting and ringer-muting circuits are discussed in the following sections.

PULSE DIALER

Recall that the pulsing switch on the rotary dial of the conventional telephone set is in series with the induction coil (hybrid) and interrupts the loop current a number of times equal to the number dialed; one time for the number 1, five times for the number 5, etc., at 10 pulses per second. This is called loop disconnect dialing. The impulse dialer electronic circuit that replaces the mechanical dial must do the same function—interrupt the loop current at 10 pulses per second.

It also must do another function. Recall that the conventional mechanical dial has additional switches to short out the handset receiver so that the dial pulses were not heard by the caller. This is called muting. The impulse dialer also must mute the receiver or, as more generally stated, mute the speech circuits. Generally, there are two ways that these functions are accomplished.

Dial Circuits in Parallel with Speech Network

The first way of accomplishing the functions is shown in the equivalent circuit of *Figure 4-3*. Here the dial circuits, represented by switch S3, are in parallel with the speech network. This previously has been called the hybrid network, but more recent terminology for this equivalent is speech network. The speech network is muted by switch S4 under control of the pulse dialer. Fundamentally, here's how it works.

When the handset is taken off-hook, S4 is closed and S3 is open. The loop current flows through the speech network in order to signal the central office that a subscriber wants service. When the caller starts to dial and depresses the first key, the pulse dialer circuit closes S3 and opens S4 so that the loop current now flows through the dial pulse circuit and at the same time the speech network is disabled so the receiver is muted. As the circuit dials, it does the equivalent of opening and closing S3 electronically to interrupt the current in the loop the required number of times. When dialing is finished, S3 remains open and S4 closes to maintain the loop current.

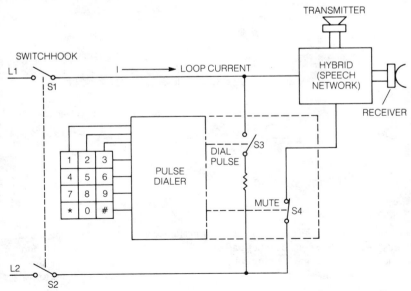

Figure 4-3. *Dial Circuits in Parallel with Speech Networks*

Dial Circuits in Series with Speech Network

The pulse dialer also may accomplish the functions of interrupting the loop current and muting as shown in *Figure 4-4*. In this case, the circuits to do the current interruption, represented equivalently by S3, are in series with the speech network. Just as in *Figure 4-3*, the pulse dialer does the equivalent of opening and closing S3 electronically. However, unlike *Figure 4-3*, where no current flowed through the speech network while dialing was going on, now the interrupted loop current is flowing through the speech network. Therefore, muting must be accomplished differently.

Electronically, the pulse dialer does the equivalent of opening S4 in the receiver circuit. Thus, even though the current is pulsing through the hybrid network, the noise does not affect the receiver because its circuit is open. Let's now look at a typical integrated circuit pulse dialer.

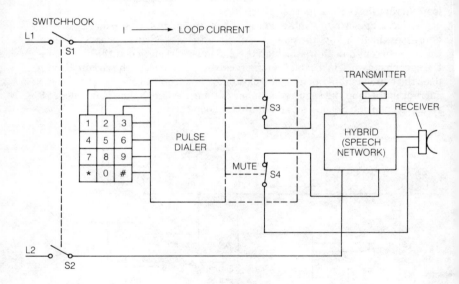

Figure 4-4. Dial Circuits in Series with Speech Networks

Primary Functions

Figure 4-5 details the primary functions contained in a typical pulse dialer integrated circuit. The keypad encoder circuit accepts inputs from four rows and three columns (or four columns for special keypads). It scans these inputs to determine when a key has been depressed. It also has a circuit to check the inputs continually over a set period of time to eliminate any error because of contact bounce.

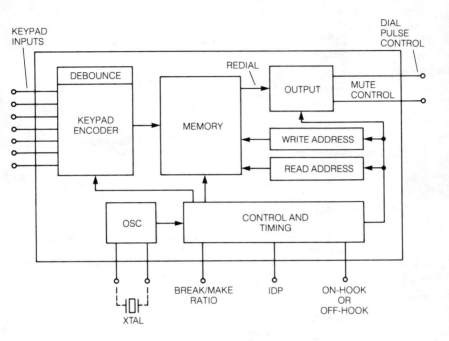

Figure 4-5. Pulse Dialer Integrated Circuit

Once detected and verified, the number of the depressed key is encoded as a binary code and stored in memory. Depending upon the particular IC, up to 17 or 20 digits can be stored in the memories provided in the integrated circuit. When the number is stored, the write address register directs the number to the correct location in memory. If the number is to be read out for redial, the read address register directs from which location in memory the number should be read.

An oscillator circuit, either controlled by a crystal or by resistors and capacitors, provides pulses at a stable frequency which determine the basic timing of the integrated circuit through counters and registers in the control and timing circuits.

The output circuitry, under the direction of the timing and control circuits, outputs control pulses for dialing and for muting. Direct pulsing of the telephone line is not practical with transistors built on the integrated circuit because the transient voltages that exist exceed the breakdown voltages available on the IC. Therefore, medium-power, high-voltage transistors external to the IC are used to pulse the line and do the muting function.

Special inputs usually provide the ability to choose the make/break ratio of approximately 60%/40% or 66.7%/33.3%, the ability to adjust the dialing pulse rate by a factor of 2:1, the ability to adjust the IDP (inter-digit pause) and the ability to sense on-hook or off-hook conditions. In some cases, these and other output pulse options are provided by mask programming.

IC Impulse Dialer Application

Several semiconductor manufacturers; e.g., TI, AMI and Mostek, offer the impulse dialer in integrated circuit form. A representative circuit is the Texas Instruments TCM1101. It is fabricated using CMOS technology and incorporates the redial function. IDP and make/break ratio are programmed with external connections; e.g. tieing B/M to V_{SS} gives a break/make ratio of 66.7%/33.3%, IDP to V_{SS} or V_{DD} gives programmed variations.

Other functions are mask programmable. Several of these are:
1. The pre-impulsing pulse (PIP)
2. Certain dialed numbers can be blocked:
    a. Numbers beginning with a particular digit; e.g., 0 or 1
    b. Numbers which exceed a certain number of digits; e.g., certain long distant numbers

In addition, further modification of the two selected values for IDP and B/M is possible by mask programming.

Transistor Muting

Figure 4-6 shows the application of the TCM1101 where the dial circuit is in parallel with the speech network. Controlling transistor Q1 does the pulse dialing; controlling transistor Q2 does the speech network muting. The DC voltage on the telephone line produces a positive voltage at terminal D of the diode bridge regardless of the polarity on the line. The line current normally flows through the bridge, through diode D1 and zener diode D2, through transistor Q2, through the speech network and through the remaining arm of the bridge back to the line. During dialing, the IC switches off Q2 with the muting signal from pin 8. With Q2 off, Q1 is turned on and off to do the line pulsing in response to the control signal on pin 7. Q1, Q2, Q3 and Q4 must have breakdown voltages of more than 200 volts.

The power for the IC is developed across the electrolytic capacitor C2. The charge on C2 is maintained for several hours after the handset goes on-hook. This charge provides power to the IC memory which has stored the last number dialed.

When a key is depressed, the IC detects the number by scanning the keypad. The timing and control, synchronized by the 455 kHz crystal, provides pulses on pin 7 to cause Q1 to be turned on and off. The timing of these pulses and the muting signal are shown in *Figure 4-7*. Note *Figure 4-7* also shows that there is an option on the type of waveform that is outputted to pin 7. This is another mask programmable feature.

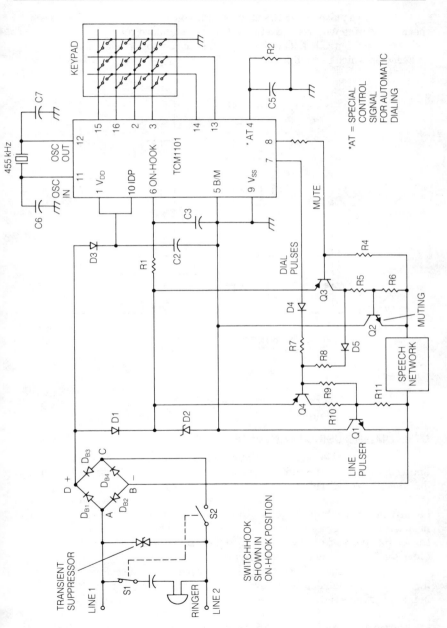

Figure 4-6. *TCM1101 Telephone Interface Using Transistor for Muting*

Some pulse dialer ICs may provide control signals to do the speech network muting with relay contacts. Either a standard relay or a bistable relay may be controlled. The choice of using transistor or relay muting usually is based on which technique costs less.

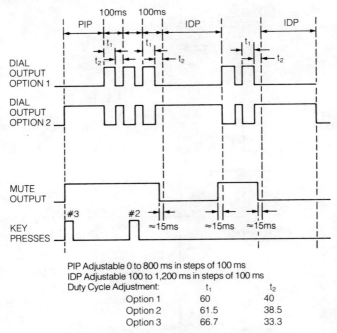

PIP Adjustable 0 to 800 ms in steps of 100 ms
IDP Adjustable 100 to 1,200 ms in steps of 100 ms

Duty Cycle Adjustment:	t_1	t_2
Option 1	60	40
Option 2	61.5	38.5
Option 3	66.7	33.3

Figure 4-7. Dial and Mute Pulse Dialing

DTMF DIALING USING INTEGRATED CIRCUITS

Dialing also may be accomplished by sending dual tones onto the line as discussed for the conventional phone. Integrated circuits have been designed to provide this function. The conventional way of accomplishing this is shown in *Figure 4-8a*. Unlike the conventional way, in which a low-frequency and high-frequency sine wave oscillator feed the speech network, the integrated circuit DTMF generator has a counter and decoder that counts pulses from a crystal-controlled oscillator and provides output codes that correspond to the low-frequency tone required and the high-frequency tone required. Each of the two outputs from the counter feed into its own digital-to-analog (D/A) converter. The D/A converter, as the name implies, converts the digital code out of the counter to a sine wave tone.

The low-frequency tone and the high-frequency tone are summed together in an operational amplifier and are fed to the speech network as a combined signal by the output stage.

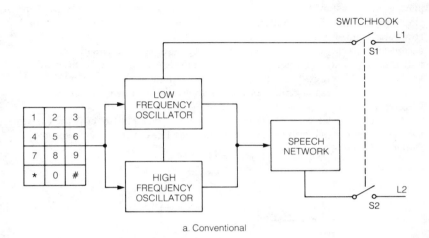

a. Conventional

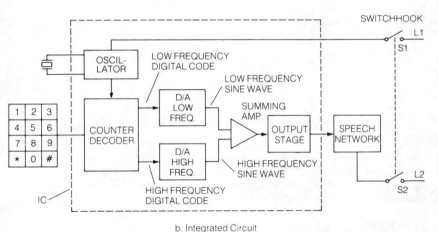

b. Integrated Circuit

Figure 4-8. DTMF Generators

An Example IC

Activation of the tone output circuits in the integrated circuit DTMF generator begins with the caller pressing a key on the keypad. The keypad contacts may be arranged as shown in *Figure 4-9*. The top schematic is a representation of a standard "2 of 8 DTMF" keypad (like a double-pole single-throw switch) where a separate set of two contacts give the row and the column of the key being pushed. An alternative and somewhat less expensive arrangement is shown in the bottom schematic which is a so-called Class A or single-pole single-throw configuration, with only one set of contacts for each row-column intersection. Some ICs are designed to accept both types; others accept only one type. In either case, the closed keypad contacts give an indication of the key being depressed. In some cases, the closed contacts may apply a supply voltage on the output lines for as long as the key is held down. In other cases, a ground or the common side of a power supply may be provided.

The keypad may be arranged to provide only a pulse, as is the case for a keypad that is used with a scanning sequence. The output waveforms are shown in *Figure 4-9b*.

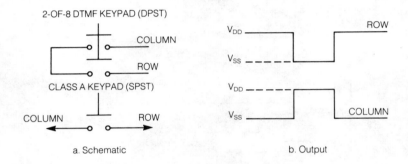

a. Schematic   b. Output

***Figure 4-9.** Keypad Contact Types and Output Waveforms*

Figure 4-10 shows more detail of a typical DTMF integrated circuit, the TCM5087 (MK5087). This circuit fits into a whole family of devices from TCM5087 to TCM5092. In the diagram, it can be seen that the decoder circuits (labeled = 1) and the set of OR and AND gates in the center have the function of controlling the output of the dividers labeled CTRDIV K. The control is such that if only one key is pushed, two outputs will be generated; one from the row dividers, and one from the column dividers. These two digital outputs are converted to analog waves and summed in the summing amplifier. Note, however, that it is possible to depress two keys on the keypad simultaneously. If the two keys depressed are in the same row or in the same column, and the Single-Tone Enable (Pin 15 on the example circuit) is a 1, only one tone, corresponding to the row or column in which the two keys are pushed, is generated. This feature is implemented to allow testing of the keypad and tone generator circuits. Operation of two keys in diagonal positions causes the circuit to provide no output at all (both sides of the decoder indicate "not equal to one").

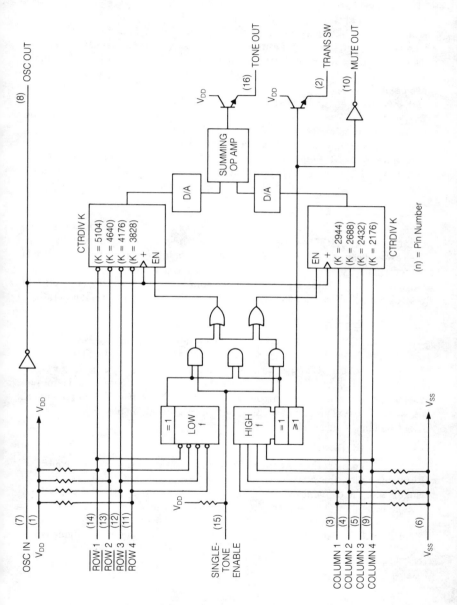

Figure 4-10. *Block diagram of TCM5087*

One other function of interest is shown by the logic element labeled ⩾ 1 at the bottom of the high frequency decoder. This gate provides an output at standard logic voltage levels (labeled Mute Out, Pin 10), and an output through a transistor switch (Pin 2), for muting of the speech circuit. The Mute Out is used to mute the receiver during dialing so the caller does not hear the full level of the dialed tones. These outputs are active whenever one or more keys are pushed.

Frequency synthesis in the circuit is done using a single input frequency of 3.579545 MHz, derived from an external crystal oscillator (using an inexpensive color TV crystal) connected to pins 7 and 8. This input frequency is divided in the blocks labeled CTRDIV K by eight different constants to produce eight possible output frequencies, two of which are selected each time a key is pressed on the associated keypad. The two tones are then added and amplified to produce the dual-tone output.

Output Waveforms

The output waveforms are not pure sine waves, but are stairstepped because of the digital-to-analog conversion. Typical stairstep approximation of the row and column sinusoidal output are shown in *Figure 4-11*. The sinusoidal waveforms are generated with a typical distortion of under 7%. The typical harmonic and intermodulation distortion content of the resulting two-tone sum is 30 dB down when referenced to the strongest column tone fundamental.

Actual Application

A TCM5087 is shown applied in a telephone set in *Figure 4-12*. The tone output feeds through the speech network to the line. A diode bridge feeds power to the IC and Z1 protects against transients. The capicator C1 maintains and filters the voltage across the IC to eliminate noise and supply variations. A mute signal controls Q1 and mutes the transmitter of the speech network. The XMIT output on pin 2 and the ST1 input on pin 15 were described in the discussion of *Figure 4-10*.

ELECTRONIC RINGER

Ringing is the way the called party is signaled that a call is waiting. Conventionally, this has been done with an electromagnetic bell. An electronic ringer, like the conventional bell, must be efficient, cheap, sturdy and reliable. Unlike bells, however, electronic ringers can provide extra features. They can sound varying pitches so that each of several telephones in an office can be identified. They can sound with increasing volume the longer the phone rings.

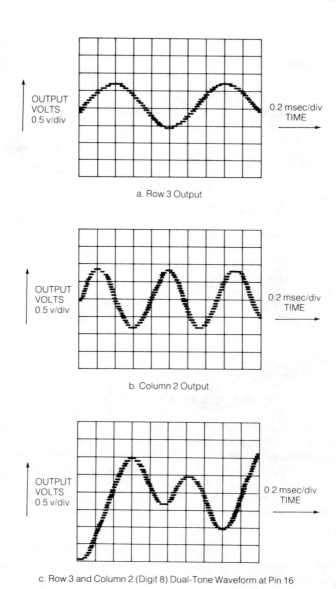

a. Row 3 Output

b. Column 2 Output

c. Row 3 and Column 2 (Digit 8) Dual-Tone Waveform at Pin 16

Figure 4-11. DTMF Waveforms

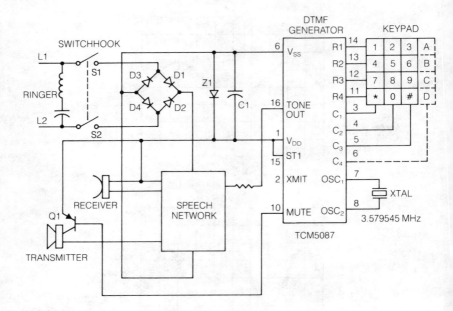

Figure 4-12. TCM5087 in Telephone Set

An important advantage provided by electronic ringers is their smaller size and lighter weight. New types of telephone sets are possible when the set does not need to contain a large heavy bell that is loud enough to be heard all over the house. Since many homes now have extension phones in one or more locations, this loudness is not needed. Component layout can be flexible so that the circuit can be fitted into unusually shaped housings or can be combined with other electronics onto a main circuit board. Electronic ringers may be either single-tone or multitone. Recall that to cause the ringing, the central office places a large amplitude ac signal on the local loop.

Single-Tone Ringer

The single-tone ringer has a fixed-frequency self-resonant oscillator which is turned on and off by alternate half-cycles of the ac ringing voltage. *Figure 4-13* shows two single-tone ringer circuits. An electromagnetic transducer is used in *Figure 4-13a* and a piezoelectric "sound disc" type transducer is used in *Figure 4-13b*.

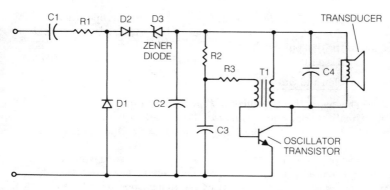

a. Electromagnetic Transducer

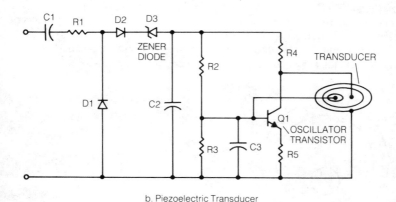

b. Piezoelectric Transducer

Figure 4-13. *Single-Tone Electronic Ringer Circuits*

Voltage Regulation

This portion of both circuits is the same. On the negative half-cycles of the ac input, D1 conducts and the power is dissipated in R1. The small voltage drop across D1 assures that D2 blocks current flow so no power is supplied to Q1 in the oscillator section; thus, there is no audio output. On the positive half-cycles, D2 conducts when the voltage rises above the zener voltage of D3 and the oscillator operates for the duration of the half-cycle. Therefore, the time of oscillation is determined by the amplitude and frequency of the ringer signal placed on the line.

Anti-Tinkle Circuit

D3 and C2 form the anti-tinkle circuit to prevent operation of the ringer by dialing pulses. The zener voltage of D3 establishes a threshold voltage that the dial pulse voltage must exceed before it can activate the ringer and C2 is a filter to absorb any transients that exceed the threshold.

Tone Generation and Output

The magnetic transducer in *Figure 4-13a* typically is a moving-coil or moving-armature transducer similar to those used for telephone receivers described in Chapter 2. Feedback for oscillation is provided by transformer coupling of T1.

The piezoelectric transducer is a thin brass disc to which a disc of piezoceramic material is glued with an epoxy cement *(Figure 4-14)*. The upper surface of the ceramic is silver plated to form one side of the electrical connection. Feedback for oscillation of the circuit of *Figure 4-13b* is obtained from a small area of the disc which is isolated from the main area by cutting or etching away the silver around the feedback element. When the circuit oscillates, the disc flexes as shown in *Figure 4-14b* because of its piezoelectric properties and produces sound. For greatest efficiency, both types of transducers must be mounted in an acoustically resonant enclosure which normally is made as part of the telephone case.

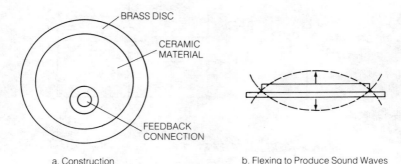

BRASS DISC

CERAMIC
MATERIAL

FEEDBACK
CONNECTION

a. Construction

b. Flexing to Produce Sound Waves

***Figure 4-14.** Piezoceramic Transducer*

Disadvantages

The single-tone ringer can be made with few components at low cost, but the output tone is usually around 2 to 3 kHz. At these frequencies, two human factors problems arise;

1. As people get older, their hearing sensitivity at these frequencies is reduced.
2. The ability of people to locate the source of a sound at 2 to 3 kHz is poor.

This latter problem increases the difficulty of determining which phone is ringing in an office with several phones of the same type on different lines. Both of these problems can be solved by using a multitone ringer at a lower frequency.

Multitone Ringer

Multitone ringers are necessarily more complex electronically than single-tone types, but with modern integrated circuits, their component count can be even lower than that of the single-tone circuits. The output signal of a multitone ringer is produced by switching between two or more frequencies at a rate determined by the tone ringer circuitry, whereas the ac ringing voltage frequency determines the switching rate of the single-tone ringer. Thus, for the multitone ringer, the ac ringing voltage's only function is to supply power to the unit.

The block diagram of *Figure 4-15* shows the circuit functions typically required in a multitone ringer. The ac ringing voltage is rectified to obtain the dc supply power which is then held constant by the voltage regulator. The overvoltage protection circuitry is the same type that has been discussed previously.

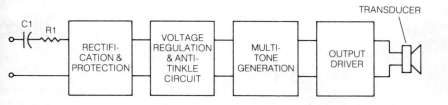

Figure 4-15. *Multitone Ringer Block Diagram*

Voltage Regulation

The ac voltage at the input to the ringer can vary over a wide range depending on how far the telephone is from the exchange. Voltage regulation is needed so that the voltage applied to the tone generation circuit is independent of the loop length. The regulator must work over a range of input voltage from about 10 to 90 Vrms. Usually the voltage regulator will provide an output higher than the minimum operating voltage (say 25 to 40 V). At lower input voltages, reduced performance is tolerated.

Anti-Tinkle Circuits

The anti-tinkle circuit must distinguish between the signal generated during dial pulsing and the ac ringing voltage. The pulses applied to the ringer during dialing are short-duration, high-voltage pulses of up to 200 V at a 10 Hz rate as shown in *Figure 4-16a*. The ac ringing voltage input is roughly sinusoidal at between 16 to 60 Hz.

One way to distinguish between the signals is to make the circuits frequency selective. The dialing transients can be suppressed by making the tone ringer respond only to the higher frequency of the ac ringing signal. The disadvantage of this method is that components for filters at these low frequencies are physically large. Another way is to reduce the sensitivity of the ringer to short pulses with a low-pass filter followed by a threshold detector. The threshold is set so that after filtering the rectified waveform, the transient signal from the dial pulses never rises above the threshold, but the ac ringing signal does, as shown in *Figure 4-16*.

In practice, this is not as easy to implement as it appears since the threshold setting also limits the level of ringing signal that will produce ringing. On long loops with more than one telephone set connected, the available ringing voltage can be as little as 10 Vrms. Therefore, some compromise may have to be made between anti-tinkle performance and ringer sensitivity.

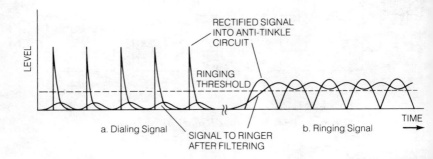

Figure 4-16. *Anti-Tinkle Circuit Waveforms*

Tone Generation

Several different techniques for tone generation are used in multitone ringers. The most common circuit produces two tones, but there are some three-tone integrated circuits available. Techniques also exist for generating more complex musical sounds.

One of the simplest schemes for generating a two-tone ringing signal is shown in *Figure 4-17a*. This uses two oscillators, one set to operate at a low frequency (e.g., 10 to 20 Hz) and the other at a higher frequency (e.g., 440 to 480 Hz). A frequency control signal switches the output between the two frequencies during ringing. This approach has the disadvantage of requiring two sets of frequency determining components which increases size and cost.

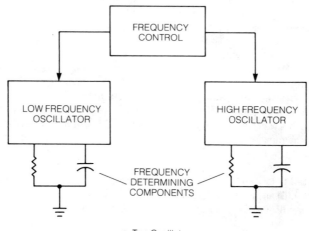

a. Two Oscillators

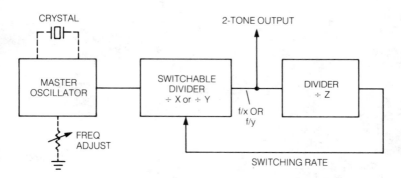

b. Frequency Division

Figure 4-17. *Multitone Ringers*

An alternative two-tone circuit uses a master oscillator to generate a high frequency from which the audio tones and the switching signal are generated by division as shown in *Figure 4-17b*. The master oscillator frequency f is divided first by X or Y to give an output of frequency f/X or f/Y Hz. This tone is divided by Z to give a switching control rate of around 10 to 20 Hz which switches the main dividers between X and Y. For most ringers, an external resistor, as shown in *Figure 4-17b*, is used to adjust the frequency of a non-crystal controlled oscillator. However, to improve frequency stability, the master oscillator in some circuits is crystal controlled as shown, but then the pitch can be varied only by changing the crystal.

Output Drivers

The output driver (power amplifier) for the ringer circuit must match the transducer impedance to the circuit to produce the maximum audio output level with the small amount of input power available. A differential output driver can be used to increase the peak-to-peak voltage drive to the transducer. This technique is especially good for driving high impedance devices such as piezoelectric transducers. Instead of the more common single-ended output shown in *Figure 4-18a*, where the transducer is connected between one active output and ground, a double-ended output circuit has the load connected between two active outputs driven differentially in opposite phase (push-pull) as shown in *Figure 4-18b*. This has the effect of doubling the peak-to-peak voltage applied to the transducer; thus, a much louder audio output is obtained.

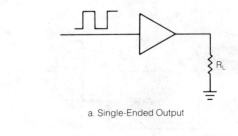

a. Single-Ended Output

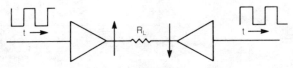

b. Double-Ended (Differential) Output

Figure 4-18. *Output Drivers*

ELECTRONIC RINGER IMPLEMENTATION EXAMPLE

Figure 4-19 shows an implementation of an IC multitone ringer that requires only two resistors, two capacitors, and the transducer external to the integrated circuit to perform the complete ringing function. All of the necessary circuitry including rectifiers and overvoltage protection is integrated into a single 8-pin dual-inline package. The operation of this circuit (a TCM1506 manufactured by Texas Instruments) is described below.

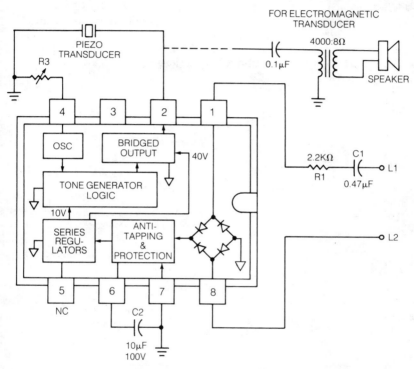

Figure 4-19. TCM1506 Ring Detector and Ringer Driver

Rectification and Protection

The ac ringing voltage passes through C1 and R1 and is rectified by the on-chip diodes. Overvoltage protection for lightning spikes up to 1500V for 200 microsecond duration is provided by a crowbar circuit. If a high-voltage pulse appears at the input to the chip, the crowbar circuit short circuits the input and dumps the excess power into resistor R1. The crowbar circuit then automatically resets. The rectified ac is filtered by an external capacitor, C2, and supplies the voltage regulator which powers the integrated circuit. A built-in oscillator supplies the timing frequency and can be adjusted with an external variable resistance, R3.

Tone Generation

The two-tone output signal is generated from a master oscillator by a programmable divider as was discussed for *Figure 4-17*. The shift rate between the high and low frequencies is controlled by another divider which counts cycles of the tone output to generate the low-frequency shift rate. Referring to *Figure 4-20*, with the master oscillator set to 48 kHz and the programmable divider set to divide-by-28, the tone output and input to the shift rate divider is 1,714 Hz. The shift rate divider (divide-by-128) counts 128 cycles at 1,714 Hz, then switches the programmable divider to divide by 32. Now the tone output and input to the shift rate divider is 1,500 Hz. The shift rate divider then counts 128 cycles at 1,500 Hz and returns the programmable divider to divide by 28. This cycle repeats as long as the ringing voltage is present on the line.

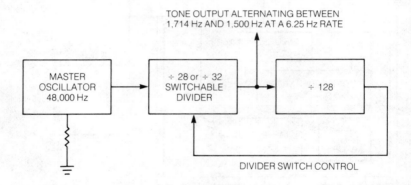

Figure 4-20. *TCM1506 Tone Generator Block Diagram*

The shift rate is given by:

$$SR = \frac{1}{(DSR\,(1/f1) + DSR\,(1/f2))} = \frac{1}{\dfrac{128}{1714} + \dfrac{128}{1500}} = 6.25 \text{ Hz}$$

where: DSR = shift divider rate ratio
                 f1 = high output frequency
                 f2 = low output frequency

The frequency of the master oscillator in the circuit is controlled by adjusting R3 in *Figure 4-19*. Since the output frequencies are generated by division from a master oscillator, they retain their harmonic relationship of 28:32 or 1:1.14 regardless of the actual oscillator frequency.

Output Driver

The output stage of the circuit is designed to drive a piezoelectric disc type transducer, or an electromagnetic transducer. The piezoelectric transducer is shown in solid lines in *Figure 4-19*. The output circuit is a single-ended design which enables the IC to generate 0 to 40 volts peak into an open circuit. If an electromagnetic transducer is to be driven, an extra capacitor of 0.1 microfarad and a transformer which matches 4,000 ohms to an 8 ohm speaker is required. The connection is shown dotted in *Figure 4-19*.

WHAT HAVE WE LEARNED?

1. Pulse dialing, tone dialing, and ring detection and sounding can be accomplished with integrated circuits.
2. Integrated circuits need special protection from pulses of voltage called transients whose magnitude is so large that it exceeds the IC breakdown voltage and can damage the circuit.
3. Voltage transients are generated when current is interrupted in an inductive circuit.
4. Pulse dialers may provide electronic switching that is either in parallel or in series with the speech network.
5. The receiver is muted so the dialing pulses are not heard by the caller.
6. Muting may be accomplished by semiconductor switching devices or by relay contacts.
7. Typical tone dialing (DTMF) integrated circuits use a master oscillator and divide the frequency to obtain the tones for the row and column matrix.
8. Both single-tone and multitone ringer ICs are available.
9. Anti-tinkle circuits prevent the ringer from being energized during dialing.
10. Multitone ringers use a master oscillator and divide the frequency to obtain a low-frequency and a high-frequency tone for the multitone ringing signal.
11. Either electromagnetic or piezoelectric sound transducers may be used for ringing.

ELECTRONIC DIALING
AND RINGING CIRCUITS

4

Quiz for Chapter 4

1. Memory is added to the telephone
set to:
   a. allow more digits to be dialed
   b. allow faster dialing
   c. allow automatic redialing of the last
   number dialed
   d. none of the above

2. What is the function of the diode
rectifier bridge in the line circuit?
   a. lower the voltage to the telephone
   electronics
   b. raise the voltage to the telephone
   electronics
   c. short out the line when the set is
   on–hook
   d. protect the set against polarity
   reversals on the line

3. What bad electrical effect happens
when the switch-hook is opened?
   a. a high voltage transient is
   generated
   b. the line is shorted out
   c. the line is opened
   d. no bad effects happen

4. Pulse dialing occurs at a rate of:
   a. 20 pulses per minute
   b. 10 pulses per minute
   c. 10 pulses per second
   d. 60 pulses per second

5. Pulse dialers must:
   a. mute the receiver in the set while
   dialing
   b. disconnect the transmitter while
   dialing
   c. short out the line while dialing
   d. all of the above

6. How many different tones may be
produced by a four-column DTMF
keypad?
   a. 2
   b. 8
   c. 4
   d. 16

7. What is an advantage of electronic
ringers?
   a. louder volume
   b. smaller size
   c. good directionality
   d. greater weight

8. What functions are provided in a
multitone ringing generator?
   a. anti-tinkle circuitry
   b. tone generation
   c. output amplifier
   d. all of the above

9. The DTMF generator in an electronic
phone produces tones using a:
   a. LC circuit
   b. RC circuit
   c. digital divider circuit
   d. digital multiplier circuit

10. A piezoelectric transducer is an:
   a. Italian whistler
   b. ceramic disk to produce sound
   c. device used as a receiver
   d. device used as a transmitter

A Microcomputer in the Telephone

ABOUT THIS CHAPTER

The preceding chapters have described the functions performed by the various components in the telephone set, and how these can be accomplished more economically (and often better) with modern electronics, especially integrated circuits. Since the microcomputer is presently the most advanced logic element in integrated circuit form, let's examine how the telephone functions can be implemented through use of a microcomputer. Then we'll see how the microcomputer makes it easy to provide "intelligence" in the telephone which opens a whole new world of useful capabilities for little or no additional cost. Providing a glimpse of that world and how to get there is the thrust of this chapter. Space does not permit a detailed discussion of how microprocessors/microcomputers operate; that topic is thoroughly covered in *Understanding Microprocessors*, another title in this series. We will, however, illustrate the design of a hypothetical computerized telephone using an actual microcomputer, the TMS1000.

WHY USE A MICROCOMPUTER?

When one hears the word "microprocessor" or "microcomputer", the natural association is with the word "computer" or "calculator". Although millions of microcomputer circuits are at work in computers and calculators, there are actually many more used as controllers in instruments, appliances, toys and games. Calculation of arithmetic functions is important in some of these applications, but a far more frequent need is for control and decision-making; in other words, the logical capability of a microcomputer is equally as valuable as its ability to add, subtract, multiply and divide.

Integrated circuits containing various combinations of logic functions have been used for years to implement electronic decision-making circuits. Increased processing and design capability have allowed semiconductor manufacturers to combine all the functions of a central processing unit (CPU) into a microprocessor and provide it in readily available quantities at low cost. Further progress has added memory and timing and control to provide a complete microcomputer in one package. The microcomputer combines programmability, decision-making, computation, memory, minimum power consumption, ruggedness and reliability with low cost. Its unique feature is programmability. Using standard hardware, the electronic equipment manufacturer has the ability to alter and upgrade equipment functions by changing a few instructions in the program (software) without rewiring the equipment (hardware). Modifications can be made to satisfy user requirements without significant interruption of user service.

Such capability is an obvious advantage in designing and manufacturing electronic equipment, but the electronic capability also is causing revolutionary change in equipment and the ways things are done outside the world of electronics.

Why Use a Microcomputer in a Telephone Set?

The use of computers in telephone switching offices has brought about a dramatic reduction in hardware size and cost, and an even more dramatic increase in efficiency and in the kinds of services offered by the telephone company. Telephone switching equipment consists of large, expensive machines that perform many simultaneous operations of detecting, searching, storing, and selecting to serve many thousands of telephones, so it is easy to see why something with the capabilities of a microprocessor might be useful for them.

It is not quite so obvious how a computer, or more correctly, a microcomputer can be used effectively in a telephone set because many of the telephone functions, as discussed in previous chapters, are simply electrical in nature and require no logic or computation. Examples of these are power supplies and regulators, amplifiers, oscillators, and the transmitter and receiver. Functions such as sending the address of the called telephone to the central office, whether by dial pulses or tones, can be accomplished by special integrated circuits such as those described in Chapter 4. So why use a microcomputer? Because a microcomputer in a telephone set can provide special features and capabilities as well as standard functions at an overall cost that is competitive with standard, no-frills telephones.

FEATURE TELEPHONES

There is a growing demand for telephone sets which provide capabilities beyond the standard ones of receiving and transmitting sound, ringing, and dialing. These are known as comfort telephones, intelligent telephones, smart telephones, or feature telephones. The latter term will be used in this chapter to describe an instrument which is equipped with an enhanced capability. The number and level of capability of these features determines whether the telephone is a "comfort" telephone with minimal extra features or an "intelligent" telephone with many features.

Comfort Telephones

A comfort telephone operates the same as the conventional telephone with features which make for easier or more comfortable use of the usual functions of dialing, transmitting, and receiving. Typical comfort telephone features are last number redial, which was discussed in Chapter 4; speed dialing, where the telephone stores a list of frequently-called or emergency numbers which can be dialed with only one or two digits; and an amplifier unit where a loudspeaker and microphone are provided for hands-free conversation.

Intelligent Telephones

An intelligent telephone has much more. It not only has some or all of the above features, many times with expanded capability, such as a larger memory to hold more speed dial numbers, but also it has features that most people might not normally associate with a telephone. These could include such things as a visual digital display, a time-of-day clock, an elapsed time indicator for the current call, an alarm for call-backs, a built-in calculator with the ability to compute the approximate cost of a toll call, automatic redialing of busy numbers (an interesting feature, since it implies the ability to detect the various busy tones from the network), automatic answering and message recording, and other features limited only by the designer's imagination.

Obviously, the definition of basic, comfort, and intelligent telephones is a matter of technology as well as the user's needs and habits. As solid-state electronics makes it inexpensive to provide these and more features, the standard telephone of the future will probably be more advanced than any comfort telephone of today.

A MICROCOMPUTER FOR TELEPHONE SET APPLICATIONS

When looking at the keypad of the pushbutton telephone, notice its similarity to a four-function pocket calculator even though the key layout is different. In fact, there are a number of similarities between the function of a microcomputer in a calculator and in a telephone set. The inputs to both are keyed in by humans operating at (to the electronics) a rather low speed and at a varying rate so a relatively slow and inexpensive microcomputer can be used. Key contacts must be debounced in both applications. As for simple calculators, decoding the keypad requires only four bits for up to 16 keys which is enough for either DTMF or pulse dialing with some keys left over for special functions.

Because we wish to power the telephone set from the telephone line where possible, the IC chip fabrication process should allow proper operation under wide variations of supply voltage. This can be obtained by using CMOS devices similar to those described in the discussion on dialers. Therefore, let's consider a low-cost and relatively simple four-bit microcomputer with the above characteristics in the design of a telephone set.

Using Standard Microcomputers for Custom Designs

Before discussing microcomputer implementation of telephone functions and features, it will be helpful to review why the discussion is based on a standard microcomputer integrated circuit. At first glance, it might seem better to custom-design a microcomputer chip for telephone set applications. However, the time required to write a program for a standard "off-the-shelf" microcomputer is usually much less than the time required to design a specialized integrated circuit, especially one as complex as a microcomputer.

Use of a software development system, usually available from the microcomputer manufacturer, makes the task of writing software less difficult and more efficient. Equally important, the resulting standard microcomputer-based design can be developed and evaluated much more rapidly with simulators, which use programs to model the function, and with emulators, which use the actual microcomputer in a special instrument for checking the design. When the program has been checked and the decision made to manufacture the system, then the final version can be mask-programmed onto the standard microcomputer circuit. This results in a permanent program that cannot be erased or destroyed; equally important, the circuit is customized for the application at minimum cost. This is because only the final metalization mask is changed from one version of the microcomputer to another; thus, the semiconductor manufacturer can mass-produce one basic chip design using the same production facilities.

The high manufacturing volumes resulting from use of a standard chip design also offer the important benefit of reliability to telephone equipment designers. The reliability of any new electronic design can be determined only by testing the final production version. With customized implementations of standard microcomputer chips, the necessary component reliability has already been proven before software is begun. The primary testing problem is only to verify the operation of the program under differing operating conditions and much of this can be done by simulation and emulation. Thus, production of equipment can begin almost as soon as the first functional microcomputer chips are received from the manufacturer.

Therefore, a standard microcomputer IC that is mask-programmed allows a new design to be done relatively quickly using a software development system, evaluated rapidly in the laboratory and in the field using emulators and simulators, and produced with a high degree of confidence in the reliability of the system.

The TMS1000 Family of 4-Bit Microcomputers

The Texas Instruments TMS1000 family of 4-bit microcomputers is typical of the general-purpose microcomputer which may be customized for a wide variety of mass production applications. Because they are designed for use in instrumentation and consumer products, many functions such as scanning the keypad, converting the outputs to various codes for simpler interface, and output drivers for display interface are built in. Derivatives of the TMS1000 are used in Texas Instruments calculators, and in other consumer products such as the *Speak & Spell*™ learning aid, electronic games, microwave ovens, food blenders, and many others.

™Speak & Spell is a trademark of Texas Instruments Incorporated

All models of the TMS1000 series have a 4-bit microprocessor and all have the same basic architecture shown in *Figure 5-1*. The interconnecting lines with the numbers means that the connection is a multiwire bus with the indicated number of wires. Different models in the TMS1000 family have various combinations of memory and input/output capability as shown in *Table 5-1*. Different semiconductor technologies; PMOS, (standard p-channel MOS), LPMOS (low voltage p-channel MOS), or CMOS (complementary MOS), are used to manufacture the family to provide for differing applications.

The CMOS versions of the TMS1000 are well suited for use in telephone sets because of the low-voltage operation, low power consumption, high noise immunity and excellent switching performance. They will operate reliably with as little as 3 volts from a power supply and still maintain computing speed. The PMOS devices are used where high-current displays or other devices need to be controlled. Each of the outputs are capable of sinking 14 to 28 mA for direct segment drive of light-emitting diode (LED) displays. Other types such as the XX70 series which allow output collector voltages to go to 35 volts, can directly drive high-voltage fluorescent displays.

Another advantage of using a family of microcomputers with the same architecture is that the same design and software may be used whether the instrument is powered directly from the telephone line (using the CMOS low-power device) or from a power supply derived from household ac power (using the PMOS device with greater display capability).

Characteristics of the TMS1000C

Let's look at more of the characteristics of the TMS1000C and 1100C, the CMOS members of the family that are well suited for telephone set applications. A functional diagram is shown in *Figure 5-2*. The TMS1000C has a memory capacity of 1024 eight-bit bytes of read-only memory (ROM) for program and constant data storage, and 64 four-bit words of static random access memory (RAM) for storage of intermediate results, telephone numbers, and similar temporary data. The TMS1100C is similar in its block diagram (architecture), but has twice as much ROM (2048 bytes) and RAM (128 words) as the TMS1000C.

Processor operation may be halted without loss of memory contents through use of the "halt" connection pin. In the halt mode, the power consumption of either type is less than one microampere. This feature is ideal for a telephone set, since the processor may be halted when the telephone is on-hook and not in use. The power consumed during normal operation is proportional to clock frequency; thus, reduction in current and voltage requirements can be accomplished by operating the processor at less than maximum speed. The minimum operating voltage is 3 volts.

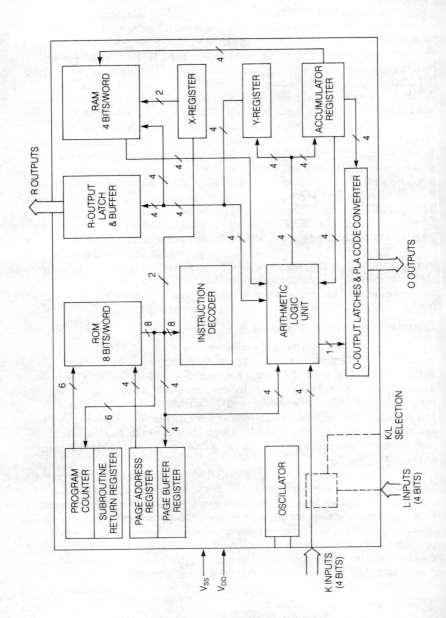

Figure 5-1. TMS 1000/1200 Block Diagram

Table 5-1. *TMS1000 Family Characteristics*

PART NUMBER	10XX	11XX	12XX	13XX	14XX	16XX	17XX
ROM 512 × 8 bits							P
1024 × 8 bits	PLC*		PLC				
2068 × 8 bits		PLC		PLC			
4096 × 8 bits					L	L	
RAM 64 × 4 bits	PLC		PLC			L	P
128 × 4 bits		PLC		PLC	L		
OUTPUTS Port 1: 9 bit addressed							P
10 bit addressed	C						
11 bit addressed	PL	PL			L		
13 bit addressed			PL				
16 bit addressed			C	PL		L	
Port 2: 8 word addressed	PLC	PL	PLC	PL	L	L	P
10 word addressed			P				
INPUTS Port 3: 4 level sensitive	PLC	PL	PL	PL	L		P
4 level + 4 edge sensitive			C			L	
SUBROUTINES 1 level, same page	PL	PL	PL	PL			P
3 level, any page	C		C		L	L	
OSCILLATOR 300 kHz	PLC	PL	PLC	PL			
600 kHz	PLC	PLC	PLC	PLC	L	L	P
1000 kHz	C		C				
SUPPLY 3-5V	C	C	C	C			
9V	L	L	L	L			
15V	P	P	P	P	L	L	P
I/O LEVELS SUPPLY (XX00)	PLC	PLC	PLC	PLC	L	L	P
35V (XX70)	PL	P	PL	P	L	L	P
PACKAGE 18 Pin							P
28 Pin	PLC	PL			L		P
40 Pin			PLC	PL		L	P

*P = PMOS, L = Low-Voltage PMOS, C = CMOS

The TMS1000 family has internal data paths that are four bits wide; that is, four bits are transferred simultaneously. Similarly, the arithmetic and logic unit (ALU) operates on four bits at once. These are the reasons why it is called a four-bit microcomputer. The instruction set is designed to handle keyed input data efficiently, with the four K inputs selectively placed directly onto the data bus. Particular family devices also have 4-bit L inputs which are latched inputs. The K/L select pin input then determines whether the K inputs or the L inputs are accepted.

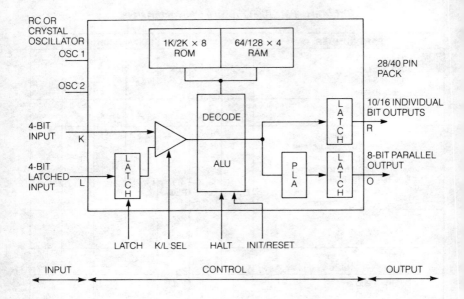

Figure 5-2. *Functional Schematic of TMS1XXX CMOS Microcomputer*

Two types of output are available; single-bit or multiple-bit. Depending on whether the 28-pin or 40-pin package is selected, either 10 or 16 single-bit output latches (designated "R" in *Figure 5-2*), respectively, may be set and cleared individually. The eight "O" outputs are latched all at the same time by one data transfer. Note that a programmable logic array (PLA) is placed between the data bus and the "O" latches; this allows four bits of data and one status bit to provide 32 different values on the "O" outputs. A PLA is a means of allowing the equipment or systems designer to choose (program) the type of code that will appear at the output for a given type of input. The PLA decoding logic is defined by the program designer and is contained in the photographic masks that are used as the integrated circuit is manufactured. This is called "mask-programming". In a telephone application, the PLA can be used to decode outputs so they will drive display devices that have seven segments or provide DTMF encoder signals directly from the binary code on the data bus inside the microprocessor.

THE MICROCOMPUTER-CONTROLLED TELEPHONE SET

A block diagram of a simple telephone instrument with a microcomputer is shown in *Figure 5-3*. Here, the microcomputer converts the keypad inputs into properly coded and correctly timed inputs to either high-voltage pulse transistors for dial pulse signaling or to a DTMF encoder chip like the TCM5087 discussed in Chapter 4. Note that the TMS1000 generates the standard row/column inputs for the DTMF chip. The dial select option, which in this case is a switch contact to an input, will enable either a pulse or a tone program, both of which are stored in the TMS1000.

For telephone sets that may be used in several countries, this method also provides a way to adapt to local signaling practices. The type of dialing is the obvious difference, but within these two categories there are many additional parameters which vary from one country to another. Because the parameters that determine the signaling frequencies, pulse rates, make/break ratios, pulse timing, and so forth are represented by values stored in the microcomputer memory, the various combinations of signaling characteristics can be implemented through software. These characteristics, instead of being programmed at the factory, could be selected by option links (wire connections) when installed or serviced. Therefore, the most common or standard configurations can be programmed on one universal standard chip,

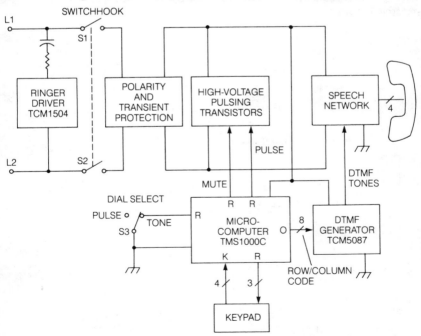

Figure 5-3. Basic Microcomputer Telephone with Dial Select

and the ones desired selected by connecting the option links. The result is that a simple, low-cost microcomputer replaces a custom chip and provides a more versatile design. Even if the microcomputer costs a little more than a custom chip, the extra cost is offset by savings through high-volume manufacturing of a standard hardware configuration.

An Enhanced Microcomputer Telephone

In *Figure 5-4*, the simple telephone set has been enhanced with several additional features, all of which are peripheral functions controlled by the microcomputer. This illustrates the principal advantage of using microcomputers in such products. The added features are essentially input and output functions of the microcomputer, and can be controlled by different programs to provide many options and adaptations for differing installation requirements. In fact, it may not even be necessary to have a different microcomputer with a different unique program for each application. If there is enough read-only memory available in the microcomputer to allow a number of different feature options to be designed into the chip, the desired features can be selected by programming the programmable read-only memory (PROM) after manufacture or at the time of installation. The approach to powering the microprocessor, scanning the keypad and control keys, and driving displays and other outputs is essentially the same regardless of selected features.

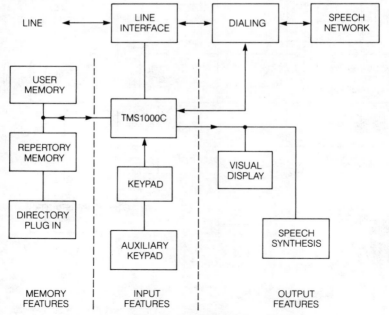

Figure 5-4. *Enhanced Microcomputer Telephone*

HARDWARE-SOFTWARE TRADEOFFS IN TELEPHONE SET DESIGN

In the design of systems using single-chip microcomputers, the software (steps in the program) and hardware are so closely related that the two must be considered simultaneously and with considerable interaction. Many times a short software routine can replace a few components to reduce system cost. On the other hand, rearranging the hardware slightly may eliminate the need for many program instructions or perhaps an entire job program (subroutine).

An example of this tradeoff is the process of "debouncing" the contacts on the telephone keypad. The input circuit should respond correctly to almost any type of contact on the keys; in other words, it must not be too sensitive or require just the right touch to operate. Very rapid depressions of the same key are probably not intended by the user; thus, the input circuit should ignore all but the first depression of the same key for some length of time. Hardware debouncing could be accomplished by a Schmitt trigger circuit for each group of keys, or by a simple RC filter circuit. Both of these solutions require additional components. A subroutine in software to do the same thing can be accomplished for virtually no additional cost; however, there may be a design impact. The debounce subroutine takes up additional memory storage space in the area of memory reserved for program storage. As a result, there may not be enough memory to program a required user feature that could be included if a hardware solution for debouncing is used.

DESIGNING A MICROCOMPUTER COMFORT TELEPHONE

In order to illustrate the design process using a single-chip microcomputer, a comfort telephone with a number of advanced features will be described. The features have been selected to illustrate the different types of functions possible with a microcomputer, rather than as a recommended list of telephone features. Both software and hardware considerations are discussed.

Features of the TMS1000C Comfort Telephone

The features described below were chosen to add the maximum capability to a standard telephone at minimum cost by using the TMS1000C with its 1024 bytes of ROM and 64 words of RAM. Obviously, if a more expensive microcomputer were used with more memory, more features could be added.

There is no standard nomenclature for telephone features; therefore, the terms most commonly used in the U.S. are used here:

1. Dual Dial Signaling: The microprocessor will generate properly-timed pulses for pulse dialing or a 4 x 4-bit code as input to a tone dialing IC.
2. Last Number Redial: Provides redialing of last dial sequence of up to 21 digits with a single key depression.

3. User Speed Dialing: Allows the user to store a telephone number of up to 16 digits and later recall the number for dialing with a single key depression.

4. System Speed Dialing: Allows short code dialing of standard system numbers such as police, fire, ambulance, weather, traffic, time, call forwarding, etc.

5. Restricted Number Access: Link Programmable. Toll access restriction prevents a number beginning with either 0 or 1 from being dialed.

6. Low-Cost Keypad: Simple, single-contact keypad. Functions may be added by adding keys.

7. Line Powered: The telephone requires only 3 volts at 1 mA from the telephone line. Battery backup for memory is optional.

The Microcomputer Control Program

Although the microcomputer clock determines the timing of its operations, two important operations are asynchronous (not synchronized to the microcomputer regular cycle time). These two operations are keypad inputs and dial pulse output.

Keypad input timing and speed are determined entirely by the user; thus, they are somewhat erratic and unpredictable. Processing this input requires a significant amount of program. The dial pulse output is obviously related to the keypad input, but the characteristics of the output pulse train are determined by the standards of the system to which the telephone set is connected. These characteristics are programmed, so that the signaling requirements are exactly as specified for the system. As discussed earlier, signaling standards for several different systems may be stored, and the proper one selected by programming the microcomputer to scan option links which will identify the parameters to be used. For this reason, it is better to let the microcomputer generate the dial pulse sequence rather than use a impulse dialer IC.

If tone dialing rather than pulse dialing is selected, the microcomputer uses a program sequence (algorithm) which will output a code to the DTMF tone generator chip. Actually, it is also possible for the microcomputer program to generate the DTMF tones directly (provided the system clock frequency is high enough), but this is wasteful of the microcomputer capability. The special DTMF IC provides better precision, and using it frees memory and microcomputer time to allow more features to be added to the basic telephone. This is another example of the hardware-software tradeoffs discussed previously.

Designing the Program

There are two basic ways to approach the design of the program to control the telephone. The first method uses an interrupt technique to handle the asynchronous input and output requirements. When an interrupt technique is used, the microcomputer ignores input/output functions until a specific request for service is made by an interrupt signal to the processor. At that time, the processor services the input or output function, then returns to what it was doing before the interrupt. Since the TMS1000C does not have an interrupt capability, a second method using a more structured approach is chosen.

In the second method, the overall task required of the microcomputer system is divided into smaller subtasks or routines which can be executed to perform functions as they fit into the overall task as required by the complete control program. Each subtask is programmed to execute in exactly the same number of machine cycles; therefore, each requires exactly the same amount of time. As a result, timing is an integral part of the algorithm and not dependent on interrupts or the particular sequence of the subtasks.

The basic control program, as shown in *Figure 5-5*, consists of three parts: initialization, input processing, and output processing. Picking up the handset causes the microcomputer to go from the standby or halt mode to the operating mode, and resets, through hardware, internal signals inside the microcomputer to initial conditions, such as status flags and pre-defined values in registers. The program then constantly circulates through the input and output routines, performing processing as necessary, until the handset is replaced. Since the time that the program spends in executing each function is exactly the same, the time required to complete one loop around the diagram is constant and may be used as the reference time interval for all time-dependent system functions.

Overall Hardware Configuration

The objective of this particular design is a telephone set which will operate from telephone line power, have the standard conventional functions, and provide useful features at minimal additional cost. Components in addition to the microcomputer will be required to provide polarity protection and transient protection. The separate DTMF generator IC is used for tone dialing. Since the microcomputer is used to generate dial pulses, the separate impulse dialer IC is not used. However, this means that discrete high-voltage transistors are needed to interrupt the line current to send the dial pulses. More on this later; for now let's concentrate on the microprocessor operation.

The TMS1000C has sufficient ROM to store the operating program, and sufficient RAM for storage of the last number dialed, the user speed number, and temporary storage required by the program.

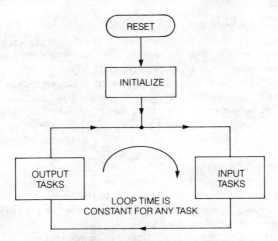

Figure 5-5. Basic Control Program

The designer must determine the keypad configuration and the output drive circuitry. The TMS1000C has four input pins (K inputs), ten individually controllable output pins (R outputs), and an eight-pin parallel output (O outputs). The function of these pins is determined by how they are connected, and their use and configuration must be established before the program can be finalized.

Input Connections

The TMS1000C has only four input pins, so the keypad must be strobed in blocks of four (or fewer) keys. This happens to be the same manner in which a calculator keypad is strobed. This works well with low-cost keypads which use a crosspoint matrix as shown in *Figure 5-6*. In this particular application, there are three types of inputs to consider:

1. The keypad digits 0-9, * and #.
2. The feature control keys: last number redial, speed dial store, user speed dial, and system speed dial.
3. The optional features: pulse dial or tone dial, and toll restriction (0 or 1).

Keypad Inputs

The first two types of input are from keys which can be implemented using a 4 by 4 keypad as shown in *Figure 5-7*. Although the illustration seems to imply a square matrix, the keypad may be manufactured with different sized and spaced keys. If all of the functions are not required, some keys may be deleted without having to change the software routine that scans the keypad. Thus, the same program can be used for everything from a standard ten-key pulse or tone dial telephone, to a full feature telephone with last number redial and speed dial functions.

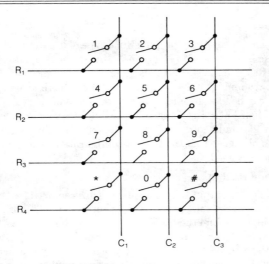

Figure 5-6. Matrix Type Keypad

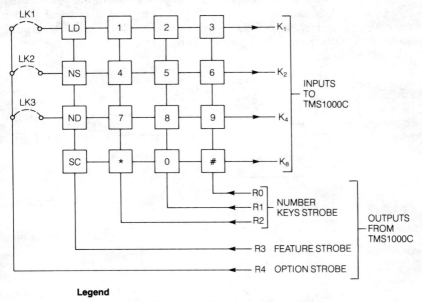

Legend

LD	Last Number Redial	LK1	Select Tone Dial Option Link
NS	Speed Dial Store	LK2	0 Toll Restrict Option Link
ND	User Speed Dial	LK3	1 Toll Restrict Option Link
SC	System Speed Dial		

Figure 5-7. Comfort Telephone Keypad/Link Configuration

Option Links

The technique for handling the third type of input, interrogation of option links, also is illustrated in *Figure 5-7*. It uses the keypad scanning technique to determine the status (present or absent) of the internal hardwired "programming" links that are selected at the time of installation. One strobe line can be used to select up to 16 different options using combinations of four links. In this case, three links are used to select the tone dialing option, and 0 and 1 toll restriction. The links are always checked during the initialization phase, so that the correct configuration is determined each time the telephone is used.

Pulse Dialing and Muting

Several pulse output timing sequences are required for the widest possible application of the telephone for different systems, but as shown in *Figure 5-8*, the impulsing method is always either of the two basic techniques discussed in Chapter 4: series impulsing and parallel impulsing. Recall that the difference is in the connection of the switch which mutes the speech circuit and in the position of the pulse dialing switch. In series impulsing, the mute switch is closed during dialing and shunts the speech network as the dial pulse switch interrupts the local loop current; in parallel impulsing, the mute switch is open during dialing and disconnects the speech network while the parallel dial pulse switch interrupts the local loop current. As discussed in Chapter 4, the mute "switch" can be a semiconductor device or a relay.

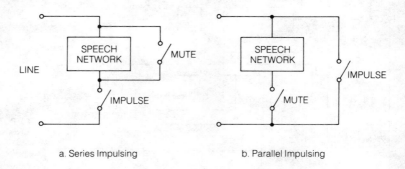

a. Series Impulsing b. Parallel Impulsing

Figure 5-8. Impulse and Muting Modes

DTMF Dialing

Most DTMF ICs are designed to operate in a standard (non-microcomputer) telephone set; therefore, they have built-in timing and debounce logic. However, in this design, the microcomputer handles the timing and debounce functions, and the DTMF chip only generates the signaling tones. *Figure 5-9* illustrates the interface between the TMS1000C and a TCM5087 tone generating circuit through use of the "O" parallel output lines. The microcomputer outputs the digit code or "key number", which is converted in the output programmed logic array (PLA) on the microcomputer chip to the row/column input expected by the TCM5087. Since the conversion logic is built into the chip, the TMS1000C software does not have to perform the conversion and the same program can drive either pulse or tone dial equipment by simply varying the PLA design and the output drivers.

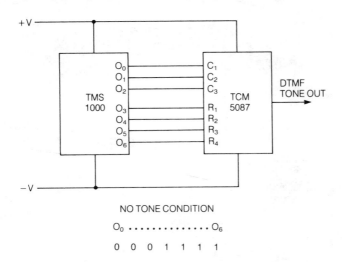

NO TONE CONDITION

O_0 ·············· O_6

0   0   0   1   1   1   1

Figure 5-9. TMS1000 to TCM5087 Interconnections

Overall Circuit Design

Figure 5-10 gives the overall circuit for the TMS1000C comfort telephone, showing the halt circuit and the rectifier bridge for supplying line power. The capacitor C1 is used to store energy to power the microcomputer during the dial pulse break period and also can provide power for a few minutes if the telephone line is interrupted when the handset is on-hook. To maintain RAM contents for a longer period, either the line power is used, or a small on-board battery, B, of the type used in watches can be built into the set and will last for several years. The switchhook contact S2 controls the halt pin of the microprocessor and also generates a pulse to initialize the program whenever the handset is lifted.

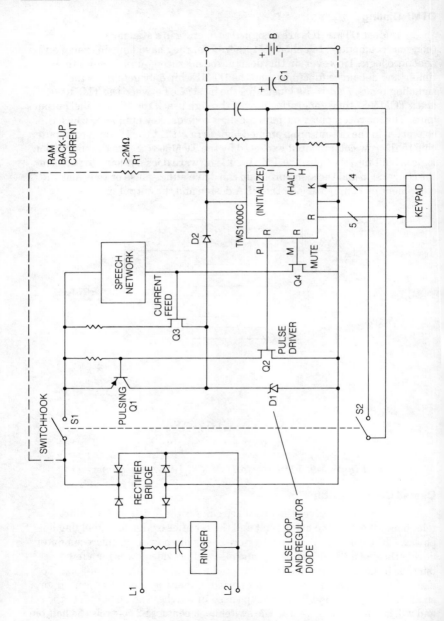

Figure 5-10. *Typical TMS1000C Application*

PROGRAM OPERATION

In any computer-based system, correct and reliable operation is dependent upon proper design of both the hardware and the software. In this relatively simple application, all functions of the microcomputer relate to dialing telephone numbers, from processing keypad inputs and storing numbers, to outputting the proper control for tone or pulse dialer circuits. The principal task is to generate a sequence of dial pulses or tones with correct timing relationships. An overall flowchart of the microcomputer program is shown in *Figure 5-11.*

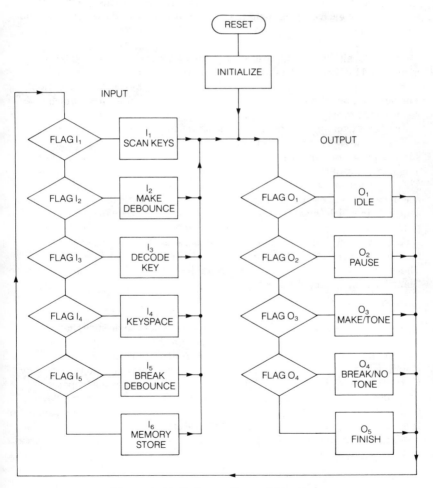

Figure 5-11. Software Control Structure

Initialization

The initialization routine is entered via the halt pin whenever the handset is lifted from the switchhook. This routine strobes the option links to determine which options are enabled and sets appropriate values in the RAM memory. It also resets the internal counters and registers, but does not alter the RAM area containing the numbers stored for speed dialing.

Timing

After initialization, the program begins a continuous loop through the input and output processing routines. The instructions are timed so that exactly half of each cycle is spent in the input portion and the other half in the output portion. This timing relationship is maintained for every sequence of operations.

Recall that the subtask subroutines were made equal so the timing is maintained by the way the program is constructed. Each instruction in a program requires a definite number of microcomputer cycles for execution. The designer determines the number of cycles for the subroutine which requires the most execution time, then inserts no-operation (no-op) instructions (commands which use execution time but do not change any values) in the shorter routines to make everything equal. In other words, each subtask program sequence time is adjusted to take the same time as the longest time required by a particular subroutine.

Each of the two sections of the program (input and output) can be in one of several states or conditions, depending on the external inputs and the sequence in which they are received. This is especially true for the input processing routine, where the programmer must not only be sure that the timing is correct, but also that every possible sequence of inputs from the user (correct or incorrect) leads to an explicitly defined operation in the program. There must be no input, no matter how unlikely, that will cause the program to "hang" in a loop without an exit.

State Diagram

The handiest tool used by software designers to be sure that there are no undefined conditions is the state diagram. *Figure 5-12* shows the state diagram for the input functions. *Figure 5-14 and 5-15* are other state diagrams. The path through the diagram (and the program) is determined by testing digital signals called software flags which are set as each operating sequence is begun and cleared as each sequence is completed. Thus, flags (either in a 1 or a 0 state) can indicate the status of the system at any particular time. If a flag is in a particular state, the system may enter a particular sequence. If the flag is in the opposite state, the sequence would be bypassed.

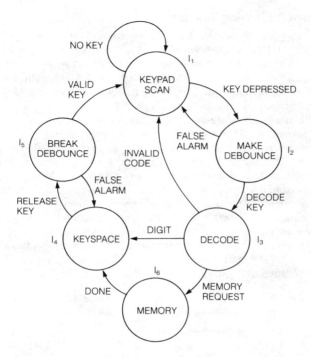

Figure 5-12. *Input State Diagram*

This method assures that correct timing is maintained regardless of
the system clock frequency; more importantly, simultaneous inputs (such as
two keys being depressed at exactly the same instant) will not cause errors in
program operation. Of course, the user may be surprised at the result, but
at least the system does not become confused and lock-up so the user
loses control.

The two factors which determine the total time required to traverse
the entire program loop are the accuracy required for output dial pulses and
the desired response time to keypad inputs. Once the longest routine has been
determined and programmed, then the minimum required total cycle time is
established. This in turn determines the minimum allowable instruction
execution time and clock frequency for the processor, since that many
instructions must be executed in one cycle. Because CMOS devices have a
current consumption proportional to operating frequency, the designer can
now determine the power supply requirements and crystal frequency for the
clock. This example application uses a clock frequency of 559.7 kHz, resulting
in an overall program cycle time of 3.3 milliseconds.

Keypad Input Processing Structure

Figure 5-12 shows that the input routine will be in one of six states, depending on the input sequence and the progress of processing as a result of the input. The input routine spends most of its time just scanning the keypad, since it has nothing else to do until a key is depressed. During this time it is in state I_1 where the strobe ouputs are sequentially set and the inputs continuously examined to detect a key depression.

When a key is depressed, the I_1 flag is reset (turned off) and the I_2 flag is set (turned on). This is so that on the next cycle through the input routine, the scan sequence of state I_1 will be bypassed and the I_2 state, "make debounce" routine, will be performed; thus, the program moves from state to state as determined by the status of the various flags. The decode state has several associated internal flags to remember the previous key inputs when handling the memory control keys and subsequent memory dialer inputs.

A sample TMS1000C assembly language routine to handle key debouncing is given in *Figure 5-13*. Each of the lines is an instruction to the TMS1000C microprocessor. The instruction mnemonic represents a short form language for what the instruction means. For example, TKA means "transfer K inputs to accumulator", and LDX means "Load register X with a constant".

INPUT 2 KEYSTROKE DEBOUNCE

THIS ROUTINE DEBOUNCES THE PREVIOUSLY DETECTED INPUT. THIS IS DONE BY CHECKING IF THE INPUT IS STILL THE SAME AS WHEN FIRST DETECTED. IF IT IS, THEN TRANSFER TO DECODE IT. IF NOT, RETURN TO KEYPAD SCANNING.

```
KDB       CALL    PAD40           PADDING

          CALL    PAD36           PADDING
          TKA                     CURRENT INPUT → ACC
          LDX     XDIGIT          > POINT TO PREVIOUSLY
          TCY     YDIGIT          > RECORDED INPUT (DIGIT)
          SAMAN                   FIND DIFFERENCE BETWEEN "THEN & NOW"
          TCY     YINFLAG         POINT TO INPUT STATE FLAGS
          RBIT    INPUT 2         RESET INPUT 2 STATE FLAG
          ALEC    0               INPUT SAME AS BEFORE (DIFFERENCE = 0)??
          BR      KDB100          YES : BRANCH TO CHANGE TO DECODE
                                  NO : GO BACK TO KEYBOARD SCAN
          SBIT    INPUT 1         SET INPUT 1 STATE FLAG
          BR      KDB999          ALL DONE—EXIT

KDB100    SBIT    INPUT 3         SET INPUT 3 STATE FLAG
          CLC                     PADDING

KDB999    BL      OUTPUT          BACK TO OUTPUT
```

Figure 5-13. *Debounce Routine Listing*

Essentially what the routine is asking the microcomputer to do is to input the signals on the K inputs, then wait a set amount of time and input the K input signals again. If the signals match, the inputs are considered valid, if not the system disregards the inputs and looks for new ones.

Keypad Decoding

Let's consider that the input system has advanced to state I_3, the decode state. The state diagram is shown in *Figure 5-14*. There are three areas where numbers can be recalled from memory in order to dial automatically. The first is controlled by the LD key, which would recall from memory the last number dialed; the second is controlled by the ND key, which would recall from memory a number that had been noted (stored) by the NS key to be saved for redialing; and the third is controlled by the SC key, which would recall from memory permanently stored numbers for weather, information, call diversion, time and temperature, etc. The particular number is located by the index digit 0-9 entered to identify which number is desired.

As shown by the state diagram, unless one of these memory keys is depressed, all other digits entered are assumed to be for dialing. In this process the input is checked to make sure the digit entered is a 0 through 9. When this is assured, the input digit is stored in the LD buffer and then the sequence is passed to the output.

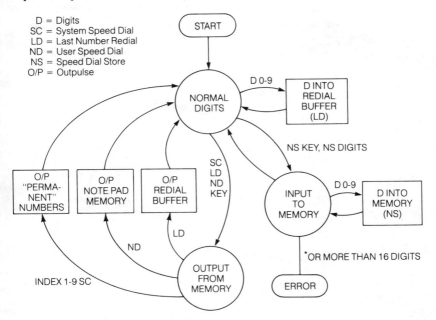

Figure 5-14. Decode State Diagram

If the NS key is pressed, the state diagram shifts to storing the number in the notepad memory. Up to 16 digits can be stored.

When a memory recall or store key is pressed, the microcomputer sets status bits (flags) to direct the program flow to store or recall the number desired.

It is possible that users will enter invalid key sequences when attempting to store or dial numbers using the memory feature. The decode routine should be designed to detect these errors and perhaps alert the user by "beeping" the ringer so that the user is not confused or frustrated in trying to use the system.

Output Signal Processing

Figure 5-15 shows the five output states and associated program flow. The output routine spends most of its time in the idle or "no output" state. A dial pulse digit output is initiated when the input section places a digit in the redial buffer in the RAM. This buffer can hold up to 21 digits. When the output section has cycled through all its states and outpulsed all the digits from the redial buffer, it returns to the idle state and waits for more output.

The telephone number remains in the redial buffer because only a "copy" of it is actually outpulsed. When the input routine detects the first key depression for the entry of a new number, the redial buffer is cleared so it can hold the new number. Note that the use of the redial buffer prevents the input routine from being slowed by the speed at which digits are dialed out by the output section.

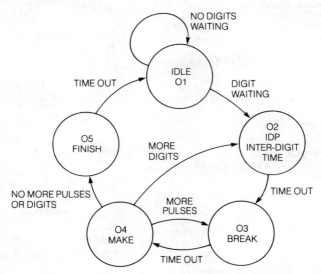

Figure 5-15. *Output State Diagram (Pulse Output)*

Timing Analysis

The performance of the system in terms of maximum throughput of telephone numbers (performance also could mean how many numbers may be stored) now may be determined by using *Figure 5-11* and a time through the input/output (I/O) cycle of 3.3 milliseconds.

Output Pulse Rate

The maximum output pulse rate occurs when one pass is made through each of blocks O3 (break) and O4 (make) of the output routine. Two passes at 3.3 milliseconds each = 6.6 milliseconds. Since frequency is the reciprocal of the period, this results in a rate of approximately 150 pulses per second at a 50% make/break ratio. This is much faster than the 10 pulses per second at a 2:1 (67%) break/make ratio required for telephone dial pulsing.

Output dialing pulses with the correct break/make ratio can be obtained by programming recurring I/O cycles with repeated passes through steps O3 and O4. As the output routine is cycled, flags are set such that 20 passes are made just with step O4 to give the break timing of the pulse and 10 passes are made just with step O3 to give the make timing of the pulse. The total of 30 cycles through the input/output routine at 3.3 milliseconds per cycle equals 99 milliseconds per pulse period or approximately 10 pps. The ratio of 20 passes to 10 passes gives the 2:1 break/make ratio.

Input Timing

Input timing is determined by the maximum allowed keying speed, which in turn depends on the minimum debounce time, since the program must accept every key depression as either a legitimate input or a bounce from the first input. Since the I/O cycle is 3.3 milliseconds, the minimum time before a key depression can be checked again after first being detected is 3.3 milliseconds. Since each state requires a pass to accomplish its function, six passes of the I/O cycle are required to progress through *Figure 5-11 and 5-12*. The time required would be 6 x 3.3ms = 19.8 ms. As a result, the fastest keying rate of input data that could be recognized is 50 digits per second. Since human input is very much slower than this, multiples of these values can be used in various combinations for the slower input to allow more debounce time for the keypad mechanism. This is accomplished by programming more passes through the input routine by using flags to repeat a particular state or states. On the other hand, if the digits were entered through a fast machine reading device, then input could be processed at the maximum rate of 50 digits per second. This might be useful for a point-of-sale terminal reading the magnetic strip on a credit card, for example.

Thus, we see that the operating speed and timing values are determined by flags and software counters which allow several options from one microcomputer program. This is another illustration of the flexibility provided by microcomputers.

Memory Allocation

Recall that both read-only memory (ROM) and random access memory (RAM) is available in the microcomputer. Read-only memory means that it is programmed and then not changed, just read. It is used to store programs and data that never changes. Random access memory, on the other hand, is frequently changed by writing into it and reading from it. It is used to store variable data.

Figure 5-16 shows the TMS1000C memory allocation for this application. The ROM contains 16 pages of 64 bytes each, which are used to store the operating program, initial values and other numbers that remain constant (system constants). The RAM contains 64 words of 4 bits each which are used to store variables and temporary data used during the task. The figure gives the approximate number of ROM pages and RAM words required for each function.

ROM		RAM	
INITIALIZE	1 PAGE	REDIAL BUFFER	22 WORDS
I/O CONTROL	1 PAGE	NOTE PAD	16 WORDS
KEY INPUT	4 PAGES	BUFFERS	10 WORDS
KEY DECODE	3 PAGES	POINTERS	6 WORDS
MEMORY CONTROL	2 PAGES	COUNTER/TIMERS	6 WORDS
OUTPUT	3 PAGES	FLAGS	4 WORDS
TIMING	1 PAGE		
NUMBER STORE	1 PAGE		

TOTAL 16 PAGES OF 64 BYTES EACH   TOTAL 64 WORDS OF 4 BITS EACH
(1024 BYTES)   (256 BITS)

Figure 5-16. *TMS1000C Memory Allocation*

Note that most of both ROM and RAM is used to implement the basic dialing operation. One page of ROM stores fixed system numbers and 38 words of RAM are for user-stored telephone numbers. The use of a microcomputer with a larger memory capacity would allow many more numbers to be stored at only a modest increase in system cost.

EVEN MORE FEATURES—AN INTELLIGENT TELEPHONE

For applications that require the capability to quickly dial numbers by using the speed dial feature, such as a business telephone, there may not be sufficient RAM available on the chosen single-chip microcomputer. Thus, a means to add external memory is desirable.

Storing and recalling such a large repertoire of numbers may be more difficult for the user, and this process can be greatly aided if a display is provided for reviewing the contents of the speed dial memory. If a display is added for this purpose, the display makes possible many other features which provide the capability for an "intelligent" telephone. Such a design will be described next, expanding from the basic design of the comfort telephone.

Add-On Memories

There are two possible types of memory that are appropriate for this application. The first is a CMOS memory using a battery for power to maintain the contents; the second is a non-volatile memory such as the Texas Instruments E^{2}PROM type. Each has its advantages and disadvantages.

CMOS Memory

An advantage of CMOS is that is can operate from the same low-voltage supply for both writing data to memory and reading it from memory. The disadvantage is that power must be supplied constantly to retain the memory contents, including situations where the unit is unplugged or the telephone line is interrupted. The power required for data retention is only 1 to 2 volts at a few microamperes, so small batteries with service lives of from two to ten years can be used.

E^{2}PROM Memory

Non-volatile memories such as the E^{2}PROM (Electrically Eraseable Programmable Read-Only Memory) do not require back-up batteries, but require, in addition to their normal low voltage operating supply, relatively high voltages (18 to 21 volts) and long memory cycle times (several milliseconds per bit) to write input data and to erase existing data. Here again is an example of an engineering tradeoff; in most consumer products, the cost of an additional power supply and the long memory cycle times may be less desirable than using a relatively inexpensive long-life battery.

Because many single-chip microcomputers do not have an external memory bus, it may seem that it would be difficult to add an external memory. However, because the additional memory is used simply as a table and handled as a peripheral device in the same way as the keypad, its connection is simple, as shown in *Figure 5-17*. A 256-word × 4-bit CMOS external memory is shown connected to a TMS1200C microcomputer.

A TMS1200C is used in this case because it has two 4-bit input ports: The K input is used for the keypad as previously described and the L input is used for the external memory input or output. This simplifies the design by separating the operations for handling the keypad from those dealing with the memory. Because the memory requires 14 control lines for addressing, read/write, chip enable, and data, some control pins on the microcomputer are shared by the memory and the keypad.

To store a digit in memory, the data word with a 4-bit pattern is placed on the four R_D pins by the microcomputer, and the address where it is to be stored is placed on the eight R_A lines used as the address bus. The R/W (read/write) line is set for writing and the EN (enable) pin is pulsed to store the data into the memory. The enable signal is a timing signal that determines the exact time that the data is written into memory.

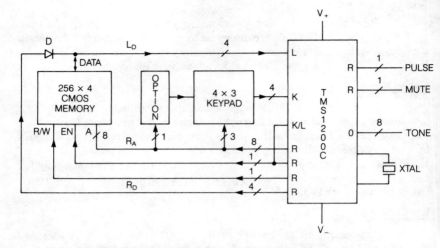

Figure 5-17. *Using External Memory with TMS1200C*

To read a digit out of the memory, the R_D pins are set to a high logic level voltage by the microcomputer, the address of the memory location from which data is to be read is put on the address bus, and the R/W line is set for reading. The output 4-bit data pattern appears on the four L input pins to the TMS1200C and are read in by the microcomputer when the ENable signal is active. The blocking diode D, isolates the 4 R_D lines from the data bus on the read cycle. The K/L input determines whether the keypad or memory input port is read. In this case, its logic state is such that the L lines are read.

Displays

When a microcomputer is included in the telephone set, it is very easy to add a display feature. For a telephone set that must operate from the telephone line, an LCD (Liquid Crystal Display) display is the best choice because it consumes very little power. However, it requires a unique multiphase drive in order to maintain a fixed ac voltage across the display. This complicates the drive interface for this type display. To simplify this interface, a special microcomputer chip, the TMS2220, has been designed.

Figure 5-18 shows a TMS2220 connected to a 6-digit, 8-segment display. The interface to the microcomputer is simple because the TMS2220, a derivative of the TMS1000, has drivers and the correct interface to drive an LCD on the microcomputer chip. The keypad inputs are as previously described. The S_{XY} outputs are the segment select signals. The COM outputs provide the necessary alternating voltage and timed signals to activate the segments in the proper phase.

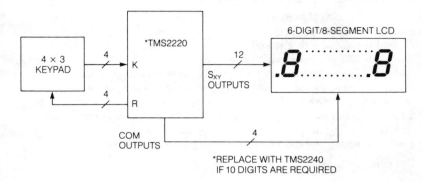

Figure 5-18. LCD Driven Directly by Microcomputer

LED Display

In telephone sets that can be powered from the ac power line, an LED (Light-Emitting Diode) display can be used. *Figure 5-19* shows the interface using discrete transistors and driver integrated circuits for the LED displays. However, some microcomputers have the driver transistors integrated onto the microcomputer chip.

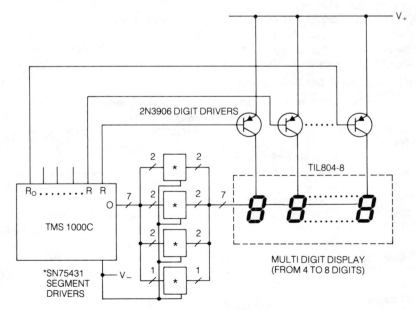

Figure 5-19. Driving Seven-Segment LED

Combining Memory and Display Functions

The use of the on-chip memory for repertory dialing requires only special keys or key sequences to indicate that speed dialing is desired. The desired speed dial code entered by the user is translated into a memory address by the microcomputer. The actual telephone number stored at that address is read from memory and sent down the telephone line by the microcomputer. Various strategies are used for number storage. A temporary buffer to display the number before dialing and/or storing is often used to avoid wrong numbers. Number storage may be made purposefully difficult, such as requiring that two keys be depressed simultaneously, to minimize accidental inputs and consequent loss of a previously stored number.

Call Charge Display

One potentially useful combination of memory and display would be to indicate the time duration and estimated cost of a toll call. Timing a call is relatively simple; it requires little more than programming the microcomputer to measure and display the elapsed time from the beginning to the end of a toll call. Since a toll call number begins with either 0 or 1, it is easy for the program to detect this event. This time will not correspond exactly to the time charged by the carrier because the carrier waits until a call is connected and answered before beginning timing, but it is close enough for estimating. The timing itself can be accomplished using a programmed loop (software method) or by an on-chip timer (hardware method).

Keeping track of the cost per minute for various types of calls and carriers is more complicated, but the concept still has value; either an average (or perhaps the highest cost) can be stored, or a table of costs for various carriers and even area codes can be stored and recalled.

The "Ultimate" Display — Synthesized Speech

When one hears or reads the word "display", one naturally assumes that a visual display is meant. This is no longer the case, however, because a verbal display is now possible. This might be especially useful in telephone applications. Since most feature telephones also contain an amplifier and loudspeaker for hands-free talking, only a speech synthesis chip and appropriate memory need to be added. Speech synthesis may provide most of the same functions as a visual display at a lower cost.

Speech synthesis ICs, such as those produced by Texas Instruments (TMS5110A, TMS5220), can produce the necessary words for a verbal display under the control of the microcomputer. *Figure 5-20* illustrates the functions required to store the phonemes for the digits 0-9 in approximately 1024 bytes of memory. This memory could be included on the synthesizer chip itself.

The speech synthesis chip is designed for simple interface to a microcomputer. The synthesizer is given an address for the start of a word and a command to speak. The synthesizer then generates speech according to the phonemes stored until an end of word code is detected. The synthesizer stops speaking and signals the microcomputer that it has finished the word. Thus, the interfacing is no more complicated than a keypad input or dialer output, requiring only a program routine to select the spoken digit. Immediate audible feedback is provided to the user as a number is keyed in, enabling him or her to verify correct key entries. Also, stored numbers can be verified by requesting that a stored number be spoken.

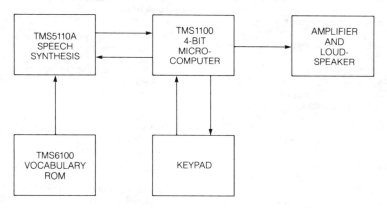

Figure 5-20. *Speech Synthesis Block Diagram*

WHAT HAVE WE LEARNED?

1. Microcomputers are self-contained integrated circuits that have input/output, memory and the central processing unit (microprocessor) all on the same chip.
2. Many extra convenience features can be had when the telephone set has a microcomputer.
3. Microcomputers can provide programmable functions that can be added to, subtracted from, or modified as the application requires.
4. The program of a microcomputer is called software. For the telephone set, it is stored in read-only memory.
5. To change functions in a telephone set that contains a microcomputer, the software can be changed rather than changing the hardware.
6. Read/write random access memory is used to store telephone numbers. If it is of the volatile type, power must be continually supplied to maintain the memory contents.
7. Dial pulses and dial pulse timing can be provided by a microcomputer.
8. Tone generation codes can be provided by a microcomputer.
9. Keypad switch contacts must be debounced to prevent improper operation. This can be done with hardware or software.
10. CMOS devices are well suited for use in the telephone set because they can operate from low voltage, have low power consumption, and have high noise immunity.

Quiz for Chapter 5

1. The purpose of putting a microcomputer in the telephone is:
  a. to provide calculator functions through the keypad.
  b. to replace wired logic and control with programmed functions.
  c. to offer games as an extra-cost option.
  d. because an electronic telephone is not possible without one.

2. A "Feature" telephone is:
  a. a model offered at a special price.
  b. only possible with a microcomputer.
  c. a telephone that offers capabilities beyond those that are standard.
  d. a telephone that provides DTMF dialing.

3. Standard microcomputers are preferred over custom designs for telephones:
  a. even though they cost more.
  b. because features can be customized, yet the microcomputer is reliable and low cost.
  c. even though they cannot do certain things, such as debouncing and DTMF signaling.
  d. even though the design takes a long time.

4. The TMS1000 microcomputer:
  a. is a single, mass-produced design that is adapted to each application by the designer.
  b. is an eight-bit device similar to the Z80.
  c. is primarily produced using TTL bipolar technology.
  d. has an interrupt capability.

5. Differences in features and signaling requirements:
  a. necessitates many different BASIC microcomputer programs.
  b. can be accommodated using a single program, using option links and/or PROMS.
  c. are very expensive to accommodate, since they require essentially a new design.
  d. are selected by the central office through the keypad.

6. The control program for the example design:
  a. consists of subtasks with precise execution times.
  b. is written in the BASIC language.
  c. is initialized when the user depresses the INIT key.

7. Inputting data from the keypad to the microcomputer is implemented by:
  a. an interrupt routine for each feature key.
  b. strobing the keys and links in blocks of four.
  c. strobing the keys, and using unique programs for each combination of links.
  d. dialing special combinations of 0-9, *, and # to access the features.

8. The input state diagram:
  a. is a stylized map of the U.S. showing terminated toll calls.
  b. leads the user through the operation of the feature telephone.
  c. is used to verify proper design of the control program.
  d. uses flags, which are set by keypad inputs to determine program flow.
  e. c. and d. above.

9. Output signal processing:
  a. refers to amplifying and filtering the transmitted speech waveform.
  b. refers entirely to the output of digits to the telephone central office.
  c. uses the same flag inputs as the input states.
  d. is only used for pulse dialing systems.

10. In choosing displays for the feature telephone, the designer would consider:
  a. that LEDs require a multiphase drive, while LCDs consume more power.
  b. that the display driver must be assembled from discrete components.
  c. that LCDs require a multiphase drive, while LEDs consume more power.

Digital Transmission Techniques

ABOUT THIS CHAPTER

This chapter explains the principles of using digital techniques to transmit telephone signals. It describes Pulse Code Modulation (PCM) and Time Division Multiplexing (TDM) systems, and compares them with analog systems. Digital transmission techniques deal with information all in digital form. To better understand the techniques used, let's cover some key concepts about digital signals.

DIGITAL SIGNALS

A digital signal, as we briefly described in chapter 1, is made up of a combination of separate parts called bits. A bit can have only two values, 0 and 1. The 0 and 1 values may represent many different two-valued conditions; on-off, full current-no current, high voltage-low voltage, true-false, etc. To understand this further, look at *Figure 6-1a*.

Single-Bit Digital Signals

The digital signal shown in *Figure 6-1a* is changing from the 0 level to the 1 level and back again as time passes. The 0 and 1 state could also be represented by voltages as shown in *Figure 6-1b*, +5 volts for the 1 level and 0 volts for the 0 level. Then the waveform of *Figure 6-1b* could be generated by measuring the voltage at switch S_o in the circuit of *Figure 6-1c* and changing the switch position from open to closed at particular intervals of time. To generate the waveform shown, S_o is held open for 1 second, closed for 1 second, open for 4 seconds, closed for 2 seconds, open for 3 seconds, closed for 1 second, open for 1 second, closed for 1 second, and open for 2 seconds.

Although this example is useful for understanding digital signals, the reader should be aware that moving switch S_0 of *Figure 6-1c* back and forth to generate the signals and measuring the digit states with meters is much too slow for real-world digital systems. Electronic circuits recognize the levels and sense and transfer the digital states between circuits at a few billionths of a second (1×10^{-9} seconds) so that millions of transactions occur in the wink of an eye.

The line L1 in *Figure 6-1c* could be a control line in any digital system, or being more specific, it could be a control line in a telephone switching network. By continually sensing the signal level on L1, the state of the control line can be determined.

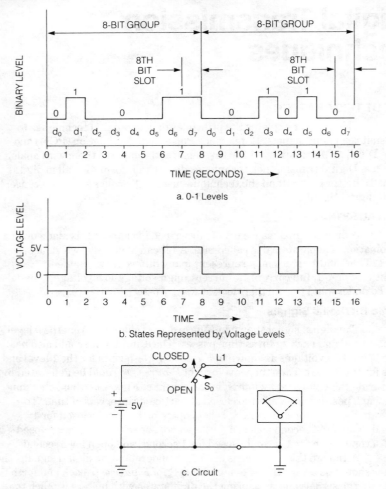

Figure 6-1. Single-Bit and Multiple-Bit Serial Digital Signals

Multiple-Bit Digital Signals—Serial

If the information contained on L1 needs to be moved or transmitted to another location, it would be transmitted in serial form. Serial transmission can be visualized by letting the waveform of digital information pass by a fixed position with time. For example, if in *Figure 6-1a* the fixed position were the Y (level) axis, and the waveform were moved to the left as time passes; for the first second a 0 level would be passing the Y axis. Then the level would change to a 1 and remain there for one second. Then the level would change to a 0 and remain at 0 for four seconds, etc. Therefore, if an observer were watching the bits go by the Y axis, the following code would be seen in the first eight seconds:

$$01000011$$

In the next eight seconds, the code would be:

$$00010100$$

Multiple-Bit Digital Signals—Parallel

Suppose now that instead of one switch as in *Figure 6-1c*, there are three switches as in *Figure 6-2a*. Each switch is switched every second as before, but the pattern of 1's and 0's that they generate is independent. However, all switching is started at the same time so every second, if a switch changes, it changes at the same time as the other switches.

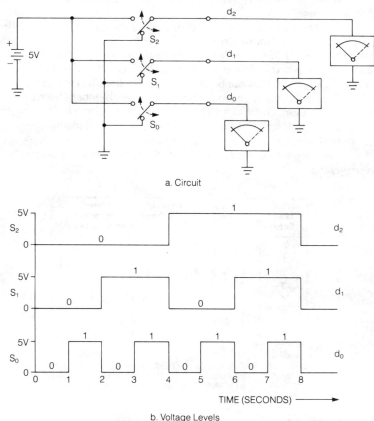

a. Circuit

b. Voltage Levels

Figure 6-2. *Parallel Digital Signals*

The 0 and 1 patterns generated by the three switches in a particular case is shown in *Figure 6-2b.* As shown in *Figure 6-2a,* these patterns were detected with a meter on each line measuring the voltage the same way as for *Figure 6-1c.* They will be used to describe another digital technique—parallel bit processing.

The patterns in *Figure 6-2b* are going to be visualized in serial form as in *Figure 6-1a.* The fixed position again is the Y axis. However, now all three waveforms are going to be moved past the Y axis at the same time. Each pattern is going to be assigned a bit position: d_0 for the S_0 pattern; d_1 for the S_1 pattern and d_2 for the S_2 pattern.

At the start (time = 0 seconds), $d_0 = 0$, $d_1 = 0$ and $d_2 = 0$. The code at time zero is 000. One second later the code changes to $d_0 = 1$, $d_1 = 0$ and $d_2 = 0$. All of the bits of the code are moving past the Y axis at the same time so the bits are appearing (being output) in parallel. This is parallel bit processing in digital systems.

The binary signals (codes) generated for the different combinations of d_0, d_1 and d_2 at the different time periods are listed in *Figure 6-3.* Column 2 shows that eight different codes can be generated using all combinations of three bits. In general, if there are N bits, there are 2^N different codes (bit patterns of 1's and 0's) that can be used to identify 2^N different conditions. As shown in *Figure 6-3,* the eight codes could represent numbers from 0 to 7, or they could represent a total of 8 different commands and/or characters. Any number of different conditions can be represented by just adding more positions to the bit pattern. An 8-bit group happens to be a convenient one used often in telephone systems. With it, 256 different conditions can be represented.

After Time (Secs)	Binary Signals			Decimal Number Equivalent	Other Possible Equivalences
	d_2	d_1	d_0		
0	0	0	0	0	A
1	0	0	1	1	STOP
2	0	1	0	2	GO
3	0	1	1	3	B
4	1	0	0	4	OFF
5	1	0	1	5	ON
6	1	1	0	6	RIGHT
7	1	1	1	7	LEFT
EIGHT DIFFERENT TIMES	EIGHT DIFFERENT SIGNAL PATTERNS			EIGHT DIFFERENT DECIMAL NUMBERS	EIGHT DIFFERENT CHARACTERS OR COMMANDS

Figure 6-3. Binary Codes Representing Different Conditions

How are Numbers Represented?

In the decimal system 10 digits are used, 0 through 9. It is called a base 10 system. In the decimal system, numbers are grouped right to left with the first digit representing the ones place (10^0), the second digit the tens place (10^1), and so on. Each place increases in value by a power of 10.

The binary system is a base 2 system. In most cases, numbers are grouped right to left the same as the decimal system. The right-most bit, d_0, is in the ones place (2^0), the second bit, d_1, is in the twos place (2^1), the third bit, d_2, is in the fours place (2^2), the fourth bit is in the eights place (2^3), and so on. Each place increases in value by a power of 2. The two systems are summarized in *Table 6-1*.

Table 6-1. *Comparison of Place Values*

	DECIMAL-Base 10				BINARY-Base 2				
PLACE	4	3	2	1	5	4	3	2	1
VALUE	1000	100	10	1	16	8	4	2	1

By using these place values wherever a 1 appears in the code of *Figure 6-3* and adding them together, the equivalent decimal number can be obtained. The evaluation of the code 011 is as follows:

$$(1 \times d_0) + (1 \times d_1) + (0 \times d_2) =$$
$$(1 \times 1) + (1 \times 2) + (0 \times 4) =$$
$$1 + 2 + 0 = 3$$

Representing Information with Binary Codes

When binary codes represent information, whether they are formed serially or in parallel, the information may be carried in a number of ways. The first way is to use the complete group of bits.

As shown in *Figure 6-1a*, a common grouping for telephone systems, as well as digital systems in general, is an 8-bit group. All of the bits in the group are needed to identify the information. The 8-bit group forms a code to identify a unique condition of the information that has been digitized. The bits in each 8-bit group are numbered from d_0 to d_7 as shown in *Figure 6-1a*; d_0 is called the least significant bit (LSB) and d_7 the most significant bit (MSB).

In addition to using the complete group of bits as a code, a second way is to use specific individual bits within a group to control specific functions or to indicate a particular status. A typical example in digital telephone systems is to assign the state of one bit in a group to indicate whether a telephone is on-hook or off-hook. An example using *Figure 6-1a* again will illustrate this technique.

Bit d_7 of an 8-bit group is used to indicate the status of the subscriber telephone set. When the handset is off-hook, d_7 equals 1; when it is on-hook, d_7 equals 0. Therefore, in *Figure 6-1a* when the first 8-bit group of bits comes by in serial form, d_7 equals 1 and the system detects that the handset is off-hook. When the second 8-bit group comes by, d_7 equals 0 and the on-hook condition is identified. Other bit positions can be used similarly to identify two conditions of a control signal.

A third way to carry information is to use subgroups of bits. Suppose the 8-bit group code is divided into subgroups as follows:

$$010 \quad 1100 \quad 1$$
$$\;\;A \qquad\;\; B \qquad\;\; C$$

In this technique, the first three bits of the code, subgroup A, could be used to identify one of eight different regional centers. Subgroup B, the next four bits could be used to identify something entirely different—one of 16 different trunk lines, for example. The last bit, subgroup C, still could be used to indicate that a handset is on-hook or off-hook.

SIGNAL CONVERSIONS

Because the early telephone systems were completely analog and a large portion of the system today is still analog, there is a need for conversion from analog signals to digital signals so they may be transmitted using digital techniques. When the signal arrives at the destination it must be converted back to its analog form. Let's talk briefly about both of these concepts.

Analog to Digital

Figure 6-4a shows the basic principle of an analog to digital (A/D) converter. The input signal is a continuously varying analog of the speech input to the telephone transmitter. At predetermined times, the input analog signal is sampled and the voltage value is converted to an equivalent digital code. In this example, an 8-bit code is used. The code comes out of the converter in parallel each time the input is sampled and its value represents the sampled voltage value. The parallel codes are converted to serial form for transmission. After arriving at the destination, they are reconverted to parallel form to feed a digital to analog (D/A) converter.

Digital to Analog

From the parallel 8-bit code input, the D/A converter outputs a voltage level for each code as shown in *Figure 6-4b*. This voltage level remains constant for the sample period and the output waveform has stepped levels. Passing the step-level output through an amplifier and filter restores the signal very nearly to its original smooth and continuously varying shape. The more samples that were taken in a given period of time by the A/D converter, the more accurate the reproduction of the signal at the output of the D/A converter.

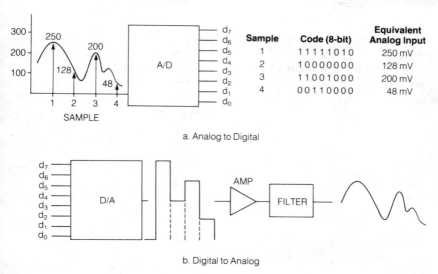

a. Analog to Digital

b. Digital to Analog

Figure 6-4. *Signal Conversions*

ADVANTAGES AND DISADVANTAGES OF USING DIGITAL SIGNALS

Changing the analog telephone network to digital has progressed rapidly since 1962, when the first digital transmission system was installed by the Bell System in Chicago. This rush to convert the analog network to digital technology has not arisen because of an overwhelming demand for transmission of data between machines, but rather because the advantages of digital transmission techniques applied to speech signals have proved to be so numerous as to make the continued installation of analog facilities a poor business choice. What are these advantages? Let's look at some of them.

Advantages

Lower Cost Using Common Circuitry

The telephone transmission network and switching equipment for digital systems can use the same kinds of integrated circuit logic as is used in digital computers. This circuitry has been declining in cost by a factor of two every three years for about the last twenty years. The cost of analog circuits has not declined as fast. As a result, digital systems cost less and as more systems are installed to increase the volume, the cost will continue to go down. There are currently no more analog switching machines being designed, except possibly very small ones. The net result is a reduced volume which has increased the cost of analog equipment. Of the currently available PBX equipment, only that offered by the Bell System is still fundamentally analog in its transmission technique.

Common Circuit Functions

When both the transmission technique and the switching system are digital, it is possible to integrate transmission and switching such that many of the traditionally required interface circuits, such as the 2-wire to 4-wire conversion shown in *Figure 6-5*, are no longer needed.

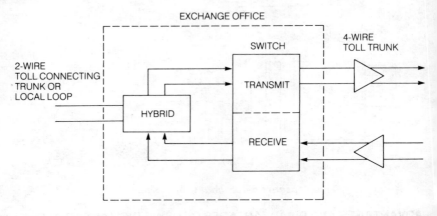

Figure 6-5. *Two-wire to Four-wire Interface for Analog Toll Trunk*

Figure 6-6 shows in greatly simplified form one end of a totally digital telephone system (telephone-to-telephone). The telephone set itself converts speech signals to digital signals. Separate circuits for the transmit and receive signals carry the signals from each telephone set to a multiplexer switch. In this example, the signals from two sets are labeled A and B, respectively. The switch multiplexes these and other signals into one continuous bit stream. (This process is explained later.) The bit stream is transmitted in serial form to a demultiplexer switch at the destination. This is shown as the receive function in *Figure 6-6*. The switch separates the individual signals from the bit stream and routes them to the proper telephone set.

The elimination of hybrids, echo suppressors, and other analog devices decreases costs and greatly increases the transmission quality for long-distance calls. In fact, if the network were digital from end-to-end, toll calls would have the same quality as local calls regardless of the distance.

Easy to Multiplex

Digital signals are easy to multiplex and the same simple low-cost digital logic circuits mentioned above can be used. The filters necessary to separate channels are much simpler than for analog multiplexing, and the need for individual wire pairs to carry the transmission is reduced significantly by multiplexing.

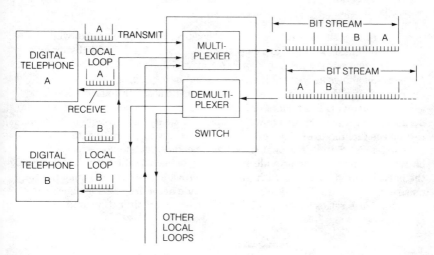

Figure 6-6. *Digital Transmission Using Multiplexing Technique*

Easier Supervision

Signaling for channel supervision and dialing is made vastly simpler and cheaper. Such signaling is inherently digital; for example, the on-hook/off-hook signal and the dial pulse train are binary 0 and 1 level signals and can be represented with simply another bit in the transmitted digital stream as shown in *Figure 6-7*.

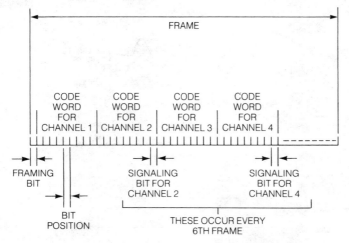

Figure 6-7. *Signaling in the Digital Bit Stream*

Better Performance in the Presence of Noise

Binary signals, because they are represented by pulses of well
defined and uniform shape, are easy to reconstruct even if badly distorted by
noise. The process is shown in *Figure 6-8* at various points in a digital system.
Figure 6-8a shows a *bipolar* bit stream. This slightly different format is called
bipolar because the 1's of the code alternate as positive or negative levels
around the 0 level. The format of *Figure 6-1a* is called a neutral bit stream
because the 1's are all positive with respect to the 0 level. In communications
terms, the technique of *Figure 6-8* gives the ability to communicate with low
signal-to-noise ratios. The analog network needs signal-to-noise ratios of
between 40 and 50 dB to provide satisfactory speech quality. This is because
the noise is amplified along with the voice signal, so only a small amount of
noise can be tolerated. Digital systems provide error-free performance with
signal-to-noise ratios as low as 15 to 25 dB with no amplification of noise.

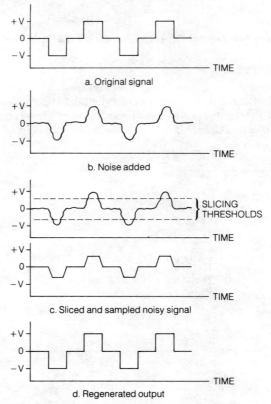

Figure 6-8. Binary Signals in the Presence of Noise

At the repeater amplifiers, shown in *Figure 6-9*, each pulse is regenerated just as it was transmitted, which eliminates any noise added along the transmission path. Since the noise is not mixed in and amplified with the signal, the speech at the end of a 2,000 mile circuit is as clear and quiet as if the circuit were only two miles long. The closer the repeaters are spaced, the lower the probability that a digital pulse will be destroyed by noise, and the higher will be the signal-to-noise ratio of the circuit. Thus, the end-to-end error rate can be made as small as desired by proper spacing of the regenerative repeaters.

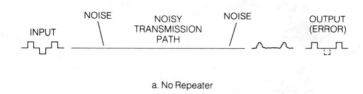

a. No Repeater

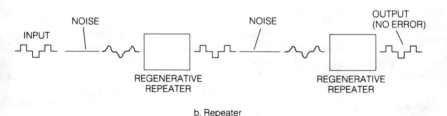

b. Repeater

Figure 6-9. *Repeater Reconstruction of Noisy Signals*

Reduced Crosstalk

Digital signals are also highly resistant to crosstalk. Crosstalk is most evident and most annoying when the two parties to a call are not talking, and can hear and understand a conversation on another circuit. Aside from being annoying, such instances breach the privacy of the parties on the other circuit. When crosstalk does appear on digital systems, it is heard as random, unintelligible noise rather than understandable speech.

Signals Can Be Mixed

Digital transmission channels readily handle digital signals from sources other than speech. *Figure 6-10* shows a string of bits from a data channel being mixed with speech that has been digitized with an A/D converter.

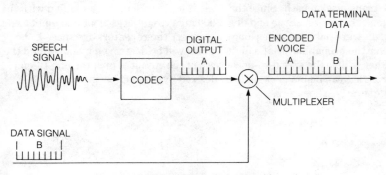

Figure 6-10. Mixing Speech and Non-speech Signals in a Digital Network

Since signals from all sources are digitized or are inherently digital, they all have the same form; thus, no special electronic tricks must be performed to keep them apart or to provide special quality channels. For the same reason, it is easy to monitor, and if necessary correct, the performance of digital channels as is done by the repeater amplifier. The monitoring circuit need only discriminate between the presence or absence of a pulse, without regard to the source or destination form of the information.

Disadvantages

As with most real-life situations, digital transmission of speech signals is not without drawbacks. Fortunately, most are related to the necessity for interfacing with the existing analog network, rather than being due to the technique or the system itself. The interface problems will become fewer as more of the network is converted to digital transmission and switching.

Extensive A to D Conversion

When a digital link must be connected to an analog network element, there is a requirement for analog-to-digital or digital-to-analog conversion. As shown in *Figure 6-11*, two such points occur at every analog office in the network. Until such time as all of these analog switching and transmission facilities are replaced, the full benefits of digital technology will not become universally available.

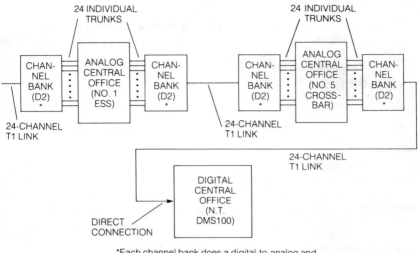

*Each channel bank does a digital-to-analog and
analog-to-digital conversion.

Figure 6-11. *A/D and D/A Conversions*

Transient Signal Environment

As mentioned above, there are advantages in using the standard
digital integrated circuits like those used in computers. However, using such
circuits in the existing telephone system also leads to a disadvantage. The
integrated circuits operate on low voltages and currents. They cannot tolerate
high voltages and their performance degrades if the temperature goes beyond
certain limits. Most of the existing telephone network (especially the central
office and subscriber loop) was designed in the era of high-current relays and
electromechanical switches. As shown in Chapter 4, such circuit conditions
produce high-voltage transients. In addition, such systems did not require
much temperature control. Consequently temperature variations are quite
extreme. The incompatibility between the two design requirements tends to
be a drawback to adding digital systems to the existing network.

Maintaining Analog System Interface

Throughout the present telephone system, the direct current to
operate the telephone and the alternating current to actuate the ringer must
be separated from the digital logic circuits in a digital system that carries
the encoded speech signals. In addition, other signal conditioning and
protection circuits must be provided. The interface circuits that provide
this separation perform functions that are known in the telephone industry
as BORSCHT. The acronym stands for the functions of Battery feed,
Overvoltage protection, Ringing, Signaling, Coding, Hybrid, and Testing.

As shown in *Figure 6-12*, all the BORSCHT functions usually are contained on one plug-in printed circuit card called a subscriber line interface card (SLIC) which is located in the digital switching facility. The SLIC provides the interface circuits to connect one subscriber line to the digital switch. These circuits will be covered in more detail in Chapter 7.

All of these functions except Coding and Testing have been described in previous chapters, but the point to be made here is that if the network were digital all the way out to the telephone set (end-to-end), a great deal of money would be saved by not having to provide all of the BORSCHT functions.

Increased Bandwidth

The most noticeable disadvantage that is directly associated with digital systems is the additional bandwidth necessary to carry the digital signal as opposed to its analog counterpart. A standard T1 transmission link (*Table 1-5*) carrying a DS-1 signal transmits 24 voice channels of about 4 kHz each. The digital transmission rate on the link is 1.544 Mbps, and the bandwidth required is about 772 kHz. Since only 96 kHz would be required to carry 24 analog channels (4 kHz × 24 channels), about eight times as much bandwidth is required to carry the 24 channels digitally (772 kHz / 96 kHz = 8.04). The extra bandwidth is effectively traded for the lower signal-to-noise ratios.

Accurate and Synchronized Timing

A technique for monitoring the state of just one bit in a group of bits was described at the beginning of the chapter. Since it is only possible to detect the presence or absence of a bit and to distinguish the meaning of one bit from another by their relationship in time, the element of timing is critical in digital transmission and switching systems. *Figure 6-13* shows how timing errors may cause bit errors in the signal. For a given switching system, the timing for all internal data transfers and transmissions is supplied by that system's internal clock. If this signal is received by a receiver running on its own clock, and that clock is running slow (as indicated in *Figure 6-13*) or fast with respect to the transmitter clock, some bits will be lost as shown in *Figure 6-13*. This occurs because the signal level present at the time of transition from 0 to 1 and 1 to 0 of the clock is the level used to recover the signal. To eliminate this problem, some receivers use a method to produce a clock signal from the digital signal itself; thus, synchronization is maintained. If several digital circuits from several sources terminate in a single receiving switch (which is normally the case), provision must be made in the receiving circuits to accommodate the small differences in sending rate of the various sources.

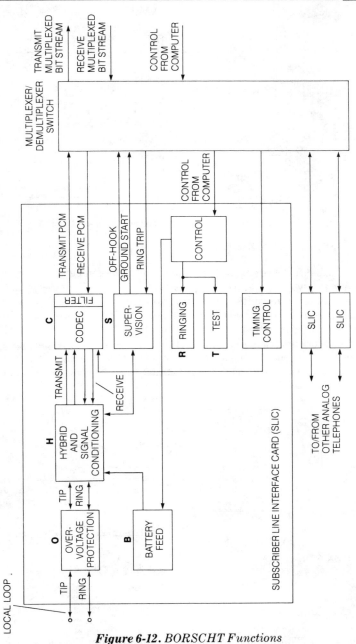

Figure 6-12. BORSCHT Functions

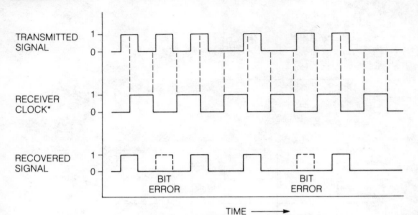

TRANSMITTED
SIGNAL

RECEIVER
CLOCK*

RECOVERED
SIGNAL

BIT
ERROR

BIT
ERROR

TIME ——▶

*Receiver clock slow with respect to transmitter clock.

Figure 6-13. *Effect of Timing Errors on Signal Recovery*

Small differences in transmission rates can be compensated for by using common low-cost IC's, but compensation for large differences would require an extensive amount of sophisticated circuitry. Large differences can be avoided by synchronizing all of the switches in a digital network, usually from a master or central timing source. This is the scheme used in the public switched network shown in *Figure 6-14*. It contains several different types of switching systems, but the master clock times all of the switching circuits in each of the systems. Now, with these advantages and disadvantages in mind, let's look at the specifics of digital transmission systems.

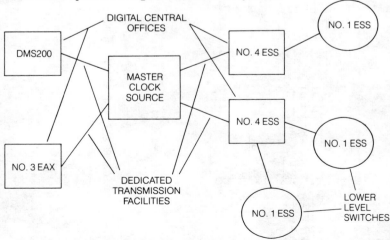

Figure 6-14. *Timing Synchronization Using a Master Timing Source*

WAVEFORM CODERS

Recall that speech signals are analog signals that vary continuously with time at frequencies over the voice band from 300 to 3,000 hertz. In order to transmit over a digital system, the analog signal must be digitized by an analog to digital converter. After the conversion, speech signals can be carried very efficiently in the form of digital pulses because the pulses necessary to represent the speech can be extremely short. As a result, the pulses can be shortened in time so that many of them can be sent in the same length of time as is occupied by the original speech. This technique is called time division multiplexing. It will be covered later in this chapter, but first, let's find out how the pulses representing the analog speech are produced.

Sampling an Analog Wave

If a person speaks into a microphone and the electrical signal generated by the input sound wave is displayed on an oscilloscope with the time base set fast enough so one-half of a cycle of the signal variation is easy to see, the waveform will appear as shown in *Figure 6-15*. The sampling times are shown superimposed on the waveform. It is apparent that the amplitude of the signal does not change very much in the short interval of time between samples. Thus, a sample of the signal at any instant in time is a close representation of the signal for a short period of time on either side of the sample point. In fact, it has been proven that if the signal is sampled at a rate which is greater than twice the highest frequency component in the signal, the samples will contain all of the information contained in the original signal. This fundamental discovery in sampling theory was made by Harry Nyquist in 1933, and is appropriately known as the Nyquist criterion. It is described by the equation:

$$f_s \geq 2\,BW$$

where f_s is the sampling frequency, and BW is the bandwidth of the input signal.

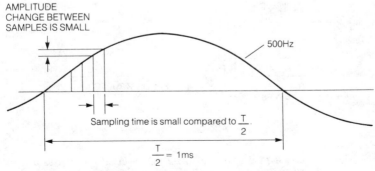

Figure 6-15. *Sampling of an Analog Waveform*

In the case of telephone speech signals, the voice channel bandwidth has been set at 4,000 Hz. Using the Nyquist criterion the sampling rate is :

$$f_s \geq 2 \times 4,000$$

which results in a minimum sampling rate of 8,000 samples per second.

Pulse Amplitude Modulation

The signal produced by this sampling consists of constant frequency pulses whose amplitude is equal to the amplitude of the sampled signal at the instant of sampling; thus, the pulses are amplitude modulated as shown in *Figure 6-16* and the process is called Pulse Amplitude Modulation (PAM). The sampled pulses can be sent on a digital channel and, when put through an appropriate filter at the output end, will reproduce the input signal as shown. However, there is a potential problem with this technique. Since the information is contained in the amplitude of the pulses, the amplitude relation of the pulses must be maintained. Any distortion or noise introduced in the transmission process cannot be taken out at the receiver, since the receiver has no way of knowing whether any particular sample value is distorted or not. In addition, another problem arises because the pulses tend to spread out during transmission and interfere with each other, making it more difficult to recover the original signal. Because of these problems, PAM is not normally used for transmission over distances greater than a few feet.

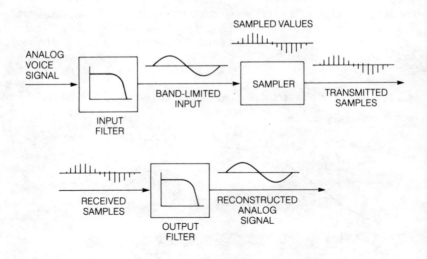

Figure 6-16. *Pulse Amplitude Modulated Transmission*

Pulse Code Modulation

To avoid the problems of PAM, the information contained in the amplitude of the signal sample is converted to a number. This is called quantization. The number then is coded into bits for transmission. Each of the bits in the code set have the same 1 level and same 0 level. The information is contained in the coded set of binary digits, not in the amplitude; therefore, the amplitude of the pulses can vary without affecting the information.

Quantization

The way a number is assigned to a particular sample is shown in *Figure 6-17*. A circuit called a quantizer takes in the sample of the analog signal and produces an equivalent number. Threshold levels are established and numbers are assigned to the samples as their amplitudes fall within the bands formed by the threshold limits. The assigned number in most cases is an approximation rather than a true value because the true value would require many more bits in the binary code. The binary code has a set number of bits which limits the unique numbers that can be assigned; therefore, the closest number to the true value is selected from the available limited set of numbers and is used to represent the sample value.

This approximation causes an error which is the difference between the approximate number and the true sample value shown as X in *Figure 6-17*.

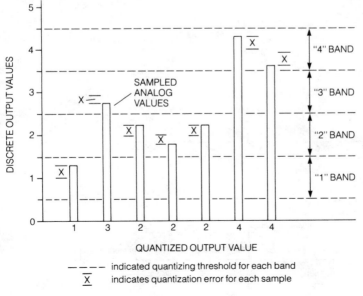

- - - - indicated quantizing threshold for each band
$\overline{X}$ indicates quantization error for each sample

Figure 6-17. *Quantization*

This quantization error adds noise to the signal, called quantization noise, which is heard on the telephone as hissing. Quantization noise can be reduced by making the threshold bands narrower. This effectively provides more intervals or numbers that can be assigned over the maximum amplitude range; thus it makes the difference between the numbers smaller to reduce the quantization error. However, providing more intervals requires more bits in the binary code; therefore, more bandwidth is required. There is a tradeoff between small quantizing intervals (higher bandwidth, lower noise), and fewer intervals (lower bandwidth, higher noise).

An effect called idle channel noise is produced in some quantizers. This occurs at small signal levels when the quantization noise is greater than the signal. This effect is particularly noticeable because there is no other signal on the channel at the time to cover up the noise. As shown in *Figure 6-18*, proper design of the quantizer results in a signal level less than the maximum value of the first quantization interval (in this case "1") being assigned a value of zero. Only the amplitudes above the first sampling threshold produce an output. This technique effectively provides an automatic squelch for small signal values and reduces the idle channel noise considerably.

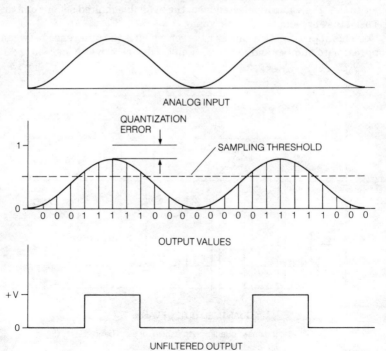

Figure 6-18. Idle Channel Noise Produced by Quantization

Coding

Once the analog signal sample has been quantized into a number, the number must be translated into a set of bits. The circuit that converts or translates the quantized signal is called an encoder, or usually just a coder. The circuit at the receiving end that performs the inverse operation (translating the bits into a number) is called a decoder. The combination of the two, which is necessary for a complete two-way system, is called a codec (from COder-DECoder). Codecs and their associated circuitry will be described in Chapter 7.

Linear Coder

The simplest form of coding produces an output which is linear with the input. A graph of input versus output is shown in *Figure 6-19*. If the input signal value is decimal 1, the coder produces an output binary number of 001. If the input value is decimal two, the output produced is the binary number 010, and so on. This scheme is easy to understand. It is used in some current business telephone systems and some of the modern digital recording systems for high-fidelity audio recordings.

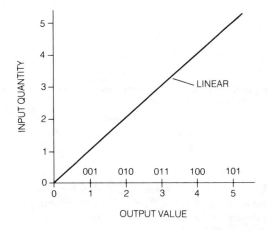

Figure 6-19. *Linear Coding*

The number of bits a coder outputs depends on the number of quantization intervals in the quantizer. The number of intervals advances by powers of 2 as bits are added in the code as shown in *Table 6-2*.

Table 6-2. *Quantizing Intervals*

Number of Bits in Code	Number of Intervals
1	2
2	4
3	8
4	16
5	32
6	64
7	128
8	256

It is necessary to quantize both positive and negative polarities of the signal; therefore, one of the bits of the code must be used for identifying the polarity. For this reason, the number of intervals is reduced by a power of 2; thus an 8-bit code would provide 128 intervals plus the sign bit. The number of bits in the code for a required number of intervals is:

$$n = \log_2 (2 \times N)$$

where: n is the number of bits
                                         N is the number of intervals

N is multiplied by 2 to gain the extra bit position for the sign bit. As an example, for 64 intervals:

$$n = \log_2 (2 \times 64)$$
$$n = \log_2 (128)$$
$$n = 7 \text{ bits}$$

Companding

One of the fundamental measures of quality for a quantizer is the signal to quantizing noise ratio, or SQR. For a linear system, the SQR is the ratio of the size of the input signal to 1/4 or 0.25 the size of a quantization interval. (This value of 0.25 for the average quantizing noise is determined statistically by assuming that, over a long period of time, the coded sample inputs have a uniform distribution of levels within a particular threshold band.) This means that the SQR increases with increasing signal amplitude, so that large signals will have a higher SQR (better quality) than small signals.

In *Figure 6-20*, a small signal of amplitude 1 has an SQR of 4 while a large signal of amplitude 5 has an SQR = 20. This condition is not desirable because the small signals are more likely to occur than large signals, and the large signals tend to mask any noise present. The remedy for this condition is to adjust the size of the quantization intervals in relation to the input signal level so that the intervals are smaller for small signals and larger for large signals. This gives a nonlinear output versus input relationship, and results in the output being compressed with respect to the input. The corresponding curve is shown in *Figure 6-21*. Note that in this figure, an input signal that increases in amplitude from 1/2 to 1 changes by 16 in its coded value, while a signal that changes from 1/64th to 1/32nd changes by the same amount, 16, in its coded output. Thus, a change in a small signal produces the same amount of output change as a signal 32 times as large.

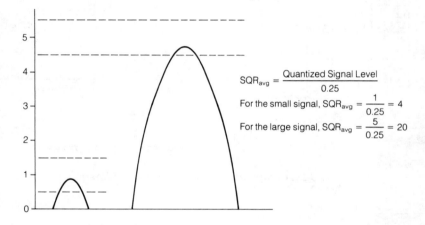

$$SQR_{avg} = \frac{\text{Quantized Signal Level}}{0.25}$$

For the small signal, $SQR_{avg} = \dfrac{1}{0.25} = 4$

For the large signal, $SQR_{avg} = \dfrac{5}{0.25} = 20$

Figure 6-20. *SQR Increases with Signal Level in a Linear Coder*

At the receive end of the transmission, the decoder has a complementary expansion characteristic to restore linearity to the signal. The combination of characteristics in the codec is called a compander (for COMpressor/exPANDER). When a compander is used, the SQR is about the same across the range of input signal levels. The devices at each end of the digital transmission channels in the public switched telephone network that perform the sampling, quantization, and coding to transform speech signals to bits are called channel banks.

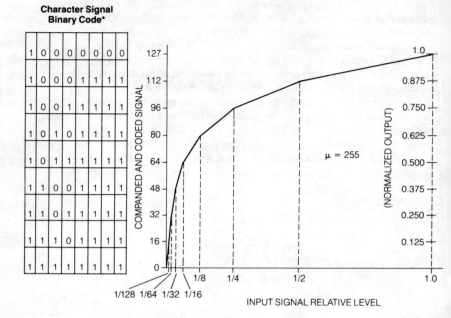

*This is the bit pattern transmitted for positive input values.
The left-most bit is a 0 for negative input values.

Figure 6-21. *Companding Curve of the μ-Law Compander*

μ-Law Compander

In a large telephone system, such as the Bell System in the U.S., large numbers of digital channels are interconnected with each other and all the channel banks must use a common scheme for the analog to digital transformations and companding required. This scheme is called the "μ = 255 law companded PCM digital coding standard", usually just called μ-law. The companding circuits operate on a logarithmic curve, using the relationship:

$$F\mu\,(x) = sgn(x)\,\frac{\ln\,(1\,+\,\mu\,|x|)}{\ln\,(1\,+\,\mu)}$$

where:

x is the normalized input signal (between −1 and 1)
sgn(x) is the sign (+ / −) of x
μ is the compression parameter, set to 255 for
the North American network
Fμ(x)is the compressed output value

The encoder operates on a segmented linear approximation to the true logarithmic curve as shown in *Figure 6-21*. The encoder produces an 8-bit output; 7 bits for magnitude plus one bit for sign. The left-most bit (most significant) is the sign bit.

The sign bit is 1 for positive input values and 0 for negative input values. The remainder of the code (7 bits) indicates the absolute value of the input signal. Since the sampling rate is 8,000 samples per second, as determined earlier, the data rate for an individual voice channel when encoded using the μ-law technique is 8,000 samples per second × 8 bits per sample = 64,000 bits per second. The transmission rate on most digital facilities is much higher than this because many channels are multiplexed together.

A-Law Compander

The μ-law companding characteristic is the standard for the North American and Japanese telephone networks. For the European network, however, the companding standard is called the A-law characteristic, whose compression characteristics are defined as:

$$F(x) = sgn(x) \frac{A|x|}{1 + \ell n(A)}$$

when
$$0 \le |x| < \frac{1}{A}$$

and
$$F(x) = sgn(x) \frac{(1 + \ell n\, A|x|)}{(1 + \ell n(A))}$$

when
$$\frac{1}{A} \le |x| \le 1$$

where:

$F(x)$ is the output compressed value
sgn(x) is the sign ($+/-$) of x
A is the compression parameter, set to 87.6 for the European network.

The A-law compander also produces eight bits per input sample in the same format as the μ-law compander and a data rate of 64,000 bits per second for each channel. Its segmented companding curve is shown in *Figure 6-22*. The A-law scheme produces a slightly better signal to noise ratio for small signals, but the μ-law scheme has lower idle channel noise. (When plotted on the scale of *Figure 6-21* and *6-22*, the curves appear the same, but plots on a larger scale would reveal slightly different curves.)

The digitizing schemes described above take enough samples and send enough bits to encode the complete input waveform so there is exact reproduction at the destination. Therefore, they are suitable for encoding and transmitting any waveform, as long as its bandwidth is limited to what the chosen sampling rate can encode without error. The capability to accurately transmit any waveform is obtained at the cost of sending enough bits to encode the entire sample at every sample interval.

**Character Signal
Binary Code***

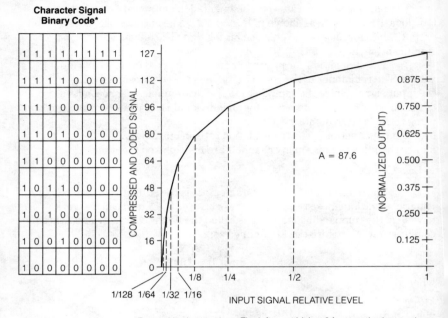

*For positive input values. The left-most bit is a 0 for negative input values. Even bits (beginning with 1 at the left) are inverted before transmission.

***Figure 6-22.** Companding Curve of the A-Law Compander*

Delta Modulation

Another technique for encoding the sampled waveform is called Delta Modulation. This scheme, rather than sending the encoded value of the sample, sends only the polarity of the difference between one sample and the next. The basic block diagram is shown in *Figure 6-23a*. A sample of the input signal is encoded. Through a feedback path, it is decoded and compared to the input at the next sample to determine if the input signal is going + or −. In essence, the output, as shown in *Figure 6-23b*, indicates only which direction the input waveform is changing, but does not indicate how fast the signal is changing. Therefore, the rate of change information must be built into the decoder or transmitted separately. In part, this problem is overcome by sampling the input at a higher rate than for the logarithmic PCM coders discussed above. This gives better sample-to-sample correlation; thus, less error in the circuits that recover the signals. At the data rates used for standard digital signal transmission, the logarithmic PCM scheme provides slightly better quality and considerably less idle channel noise. This is because a delta modulator has no way to represent a zero output and produces a spurious signal in an otherwise quiet channel as shown in *Figure 6-24*.

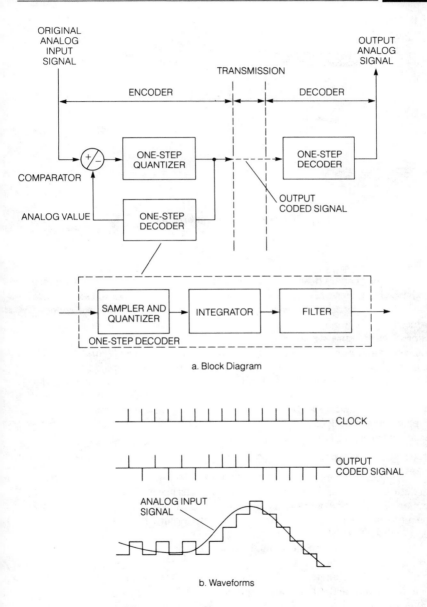

a. Block Diagram

b. Waveforms

Figure 6-23. *Simple Delta Modulation*

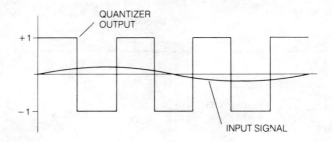

Figure 6-24. *Idle Channel Noise Generation in Delta Modulator*

Source Coders

The coders described above belong to the general class of waveform coders which tries to reproduce the waveform of the input signal exactly at the output of the decoder. There is another class of coders called source coders, which seeks to minimize the number of bits (or the bandwidth) necessary to reproduce an input signal that is intelligible, but is not necessarily a faithful reproduction. For example, a reproduced speech signal can be understood, but may sound "hollow" or like a monotone. Source coders make use of prior knowledge of the characteristics of the source of the input signal in encoding that signal. There are several techniques currently in use, the most common of which is called Linear Predictive Coding. These are used a great deal in reproducing speech electronically. They are quite detailed and beyond the scope of this book.

TIME DIVISION MULTIPLEXING

Once the input speech has been sampled, quantized, and encoded in digital form, it must be transmitted to its destination. The economics of public telephone network transmission dictate that many individual channels be multiplexed over a single large-bandwidth circuit. It is generally not economical to send only one encoded voice channel at 64,000 bits per second over a single transmission channel, although it is done in some modern PBX's using digital telephone sets. Recall from Chapter 1 (*Figure 1-15*) that for digital transmission, the method of multiplexing is to send the individual bits separated in time (rather than frequency or phase), hence the name time division multiplexing (TDM).

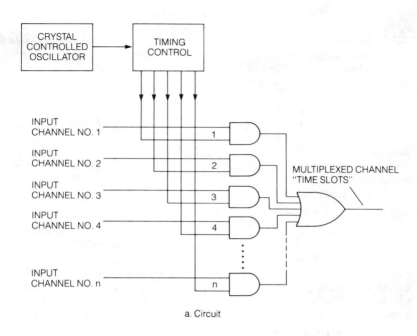

a. Circuit

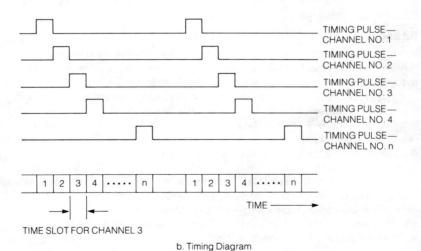

b. Timing Diagram

Figure 6-25. Basic Principles of Time Division Multiplexing

Figure 6-25 outlines the basic principles. In *Figure 6-25a*, a timing pulse generated from a master crystal controlled oscillator is one input to an AND logic gate. The digital code from a voice channel is the other input. When the timing pulse is present AND an input channel pulse is present, the signal output is fed to an OR gate which reproduces at its output any AND gate output that has been turned on by the timing pulse.

The timing pulse for each channel is produced in sequence as shown in *Figure 6-25b* producing slots of time in sequence. When timing pulse number 1 is present, a time slot is created that contains the digital code for channel 1. Following it in sequence is the time slot for channel 2, created by timing pulse number 2. Channel 3 follows channel 2, channel 4 follows channel 3, and so on up to the maximum number of channels multiplexed. The cycle then repeats. In each channel time slot, the encoded symbol appears to identify the information that came in as an original analog signal.

Synchronous and Asynchronous Systems

Assignment of a place in time in a bit stream to put the bits for an individual channel can be done on a dedicated (permanently assigned) basis, or on an as-required basis. Systems that make the assignment on a dedicated basis are called synchronous systems, and those that make the assignment on an as-required basis generally are called asynchronous systems. Depending on how they are designed, the asynchronous systems may be called either asynchronous TDMs, statistical multiplexers, or packet switches. In general, transmission systems in the present telephone network are synchronous, while networks designed especially for carrying data communications are of the asynchronous type.

Bit and Word Interleaving

Each of the digital code words from each voice channel encoder contains several bits (typically eight). Four-bit words are used in the example in *Figure 6-26*, and as shown, it is possible to interleave the bits in more than one way. If the entire code word from the first channel is sent, then the word from the second, then the third, etc., the resulting set of bits is said to be word interleaved (*Figure 6-26b*). If the bits are interleaved such that the first bit is taken from each code word in sequence, then the second bit, then the third, etc., the resulting set of bits is said to bit interleaved (*Figure 6-26c*). In either case, the resulting set of bits is called a frame.

The lowest level of TDM in the public network uses word interleaving, chiefly because the source of the data (a channel bank) produces individual code words as outputs from each digitized channel. The higher-level multiplexers in the public network are bit-interleaved because the data sources are continuous bit streams from the lower-level multiplexers.

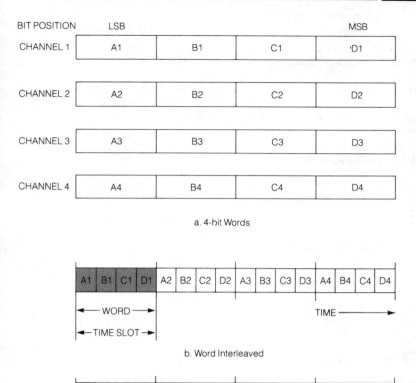

Figure 6-26. Word and Bit Interleaving

Synchronization

The sending end of the multiplexed stream of bits must add framing information to the bit stream to enable the receiving end to identify the beginning of each frame. The framing information may consist of a single bit, a code word of the same length as the others in the frame, the deletion or systematic alteration of a bit in the code words, or alteration of the electrical waveform of one or more of the bits in the frame. The schemes generally used in the telephone network add either one bit or one code word (eight bits) to the data stream to identify frame boundaries. Two of the most common are shown in *Figure 6-27*.

The T1 Transmission System

The multiplex equipment carrying the smallest number of channels in the heirarchy of digital multiplexers is called the primary multiplex. In the U.S. network, the primary multiplex signal is called the DS-1 signal, and the transmission system is called the T1 digital carrier system. As shown in *Figure 6-27a*, a frame in the T1 system consists of 193 bits. The first bit is used to establish the beginning of the frame (the framing bit). Following it are 192 bits comprising 24 code words of eight bits each. There are 8,000 samples a second and each one generates a code word; therefore, a code word occurs every 125 microseconds ($1/8 \times 10^{-3}$). The frame of 193 bits containing the 24 multiplexed channels must occur within the 125 microseconds to be able to multiplex the next sample from each channel correctly. 193 bits in 125 microseconds ($193/125 \times 10^{-6}$), is equivalent to 1,544,000 bits per second or 1.544 megabits per second (Mbps).

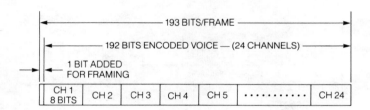

a. Added Bit Framing (DS-1 Signal Example)

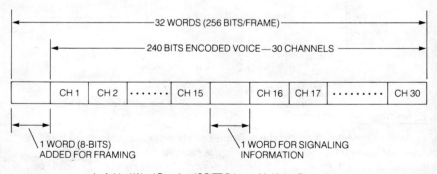

b. Added Word Framing (CCITT Primary Multiplex Example)

Figure 6-27. *Added Bit and Added Word Framing Methods*

European Systems

The European standard primary digital multiplex signal differs from the U.S. standard in that it specifies a frame with 32 code words of eight bits each as shown in *Figure 6-27b.* The first word in the frame is used for the framing information, the second through sixteenth words are for encoded speech, the seventeenth word for signaling information for each channel, and the remaining 15 words for more encoded speech channels. Thus there are 30 data words, one framing word, and one signaling word in each 32-word frame. Since the frames must occur at the standard 8,000 per second rate, the data rate is 256 bits in 125 microseconds (256 bits/125 $\times$ 10^{-6}) or 2.048 Mbps. This difference is carried through the other multiplexers in the standard heirarchy of equipment.

Pair Gain Systems

Systems that use the techniques described above to cut down on the number of wire pairs needed to carry telephone channels are sometimes called pair gain systems. The simplest pair gain system multiplexes only two conversations on a single wire pair, usually using frequency division techniques. More sophisticated systems such as the Subscriber Loop Multiplex system concentrates up to 80 subscribers on a single T1 carrier system. The Subscriber Loop Carrier 40 system can carry 40 subscribers on a 40-channel T1 line at 38,000 bits per second per channel. These systems are attractive in rural areas where the cost of providing individual wire pairs for each subscriber is prohibitive.

Analog and Digital Multiplexer Systems

Placing ever larger numbers of channels on a single transmission facility brought a necessity for a family or heirarchy of multiplex systems, both in the analog and digital domains. Refer back to *Table 1-4* which shows the North American analog multiplex systems and the transmission medium used for each channel. *Table 1-5* is a similar table of the North American digital multiplexer systems and the medium used. The European standard systems are shown in *Table 6-3.*

Table 6-3. European Multiplex Systems

Level	Number of VF Circuits	Multiplex Designation	Data Rate (Mbps)
1	30	Primary	2.048
2	120	M12	8.448
3	480	M23	34.368
4	1920	M34	139.264
5	7680	M45	565.148

Note that the European system is not integrally related to the U.S. system either in number of channels or data rate, making a rather complex interconnection problem at any level above that of the individual channel.

Data Under Voice

The Bell System has developed a special radio terminal, called the 1A-RDT, which allows the addition of a single 1.544 Mbps T1 digital channel below the lowest frequency of a 600-channel mastergroup analog multiplex carrier system. This system was developed to implement a service offering of the Bell System called Dataphone Digital Service (DDS)® for data communication. Thus, a digital capability was added to an existing analog long-haul voice carrier system.

Signaling in the PCM Word

Signaling information is carried in an analog channel by one of the several methods discussed in Chapter 1, the earliest and still most common method being that of opening and closing the dc loop. Since a PCM channel has no DC continuity, other means must be found for conveying on- and off-hook and dialing signals along with the speech. The method used in current D-2, D-3, and D-4 channel banks involves using a specified bit in a specified code word as a binary indicator of the on-hook/off-hook state of the channel. The bit used is the least significant bit in every sixth code word (*Figure 6-7*). Use of this bit which normally carries part of the speech information introduces a small amount of error or distortion into the speech signal, and lowers the SQR ratio by about 1.8 dB. A more extensive treatment of these methods is given in Chapter 8.

WHAT HAVE WE LEARNED?

1. Digital systems use information in binary form. Code words consisting of bits that have binary values 0 or 1 are used to carry information.
2. Eight-bit words are a common bit group in telephone systems. The information in the code word may be contained in the total group, in subgroups within the group or in individual bits of the group.
3. Transmission of code words in serial form is the most common method of transmitting digital signals in the telephone system.
4. Analog signals are sampled at 8,000 times per second because this is twice the maximum frequency of speech to be transmitted.
5. Analog signals are sampled, quantized and encoded into Pulse Code Modulated (PCM) binary signals. In the encoding process, the information may be compressed before transmission.
6. At the receiving end, the binary code is decoded and filtered to produce the original signal. It will be expanded if it has been compressed at the sending end.

---

[1]Dataphone Digital Service (DDS)® is a registered service mark of AT&T Co.

7. The equipment that compresses and expands the binary code is called a compander.

8. There are two common types of companders; μ-law and A-law.

9. PCM digital signals are multiplexed onto a single transmission medium by time division multiplexing (TDM).

10. Two common ways of multiplexing signals is by word interleaving or bit interleaving.

11. Multiplexed systems can be either synchronous or asynchronous.

12. Asynchronous systems are used mostly for data communications.

Quiz for Chapter 6

1. The telephone network is being converted to digital operation primarily to:
   a. carry digital computer data.
   b. reduce costs.
   c. improve speech quality.
   d. increase system capacity.

2. Digital transmission and signaling are useful because:
   a. digital logic circuits are cheaper than analog.
   b. signaling is easier.
   c. it can provide a lower signal-to-noise ratio.
   d. all of the above.

3. Digital transmission has the disadvantage of:
   a. requiring more bandwidth than analog.
   b. may not work with existing equipment.
   c. using equipment that is more susceptible to environmental extremes.
   d. all of the above.

4. The equation fs ≥ 2BW is called the:
   a. Shannon theorem.
   b. Nyquist criterion.
   c. Erlang law.
   d. Edison effect.

5. Sampling the analog wave produces:
   a. impulse noise.
   b. phase distortion.
   c. pulse amplitude modulation.
   d. frequency coherence.

6. Quantization assigns:
   a. voltages to digital signals.
   b. operators to incoming calls.
   c. numbers to analog samples.
   d. none of the above.

7. The simplest form of coding is:
   a. diphase.
   b. hybrid.
   c. compressed.
   d. linear.

8. A coder for a 64-interval quantizer must produce how many bits?
   a. 2   c. 8
   b. 10   d. 7

9. Between two quantizers, the quality of the one with the lower SQR will be:
   a. higher.
   b. lower.
   c. the same.
   d. not measurable.

10. If the code word is only 4-bits for a T1 system with a frame of 193 bits, how many channels can be transmitted?
    a. 48.
    b. 50.
    c. 6.
    d. 24.

11. In question 10, what is the bit rate in Mbps?
    a. 0.772.
    b. 3.088.
    c. 1.544.
    d. 1.024.

12. The parameter describing the degree of signal compression by the companders in the U.S. network is:
    a. sigma.
    b. μ.
    c. A.
    d. R squared.

13. In digital multiplexing systems, bit interleaving is used in:
    a. lower-level systems.
    b. higher-level systems.
    c. to interleave a code word.
    d. none of the above.
    e. all of the above.

14. Synchronous multiplexed systems have the time placement of bits:
    a. dedicated.
    b. unassigned.
    c. random.
    d. as required.

15. Asynchronous multiplexed systems are used:
    a. mostly for voice transmission.
    b. mostly for data transmission.
    c. to carry only speech information.
    d. all of the above.

Electronics in the Central Office

ABOUT THIS CHAPTER

The first five chapters concentrated on the functions of the telephone set and how the use of electronics accomplishes the functions more easily, better or with more features. Chapter 6 started the concentration on what's beyond the telephone set. This chapter continues that emphasis, taking up specifically what's on the end of the local loop—the central office. Just as the designer of an electronic telephone set must observe the existing standards and practices, so, too, the designer of the central office equipment must be sure that the circuits will respond properly to either a modern electronic telephone set or a conventional telephone set connected to the local loop. These compatibility requirements have been discussed previously; they stem from the fact that it is usually too costly to replace all of the telephones at the time a new switching office is installed, so the new switch interface with the subscriber is designed to allow the use of existing telephones.

These standard interfaces, and how to improve their operation through electronics, is the principal thrust of this chapter. There will be, however, a glimpse of the exciting world of tomorrow's local loop, when a digital telephone connects with a digital subscriber interface and utilizes the principles of the digital switching system described in Chapter 6. In addition, much of what will be covered can be applied to specialized switching systems, such as PBXs, where the telephone set can be of any design so long as the interface between the specialized system and the central office meets the telephone company standards.

THE LOCAL LOOP

The local loop, as discussed many times previously, connects the subscriber telephones with the local central office, and through the central office to the worldwide telephone network. The local loop operation has changed little since the invention of the telephone, though there have been many improvements in its construction. The principal change has been that twisted pair cables have largely replaced open wires strung on pole crossarms. The use of copper wires and relays as the primary system components was unchallenged for almost a hundred years, and they accounted for a major share of the capital and maintenance costs of the telephone industry. But beginning in the late 1960's, the economics of local loop design began to change. The cost of cable pairs continued to increase because of the copper, and especially because of the cost of installing them under the streets or on poles. At the same time, the cost of electronic "intelligence" was decreasing dramatically as integrated circuits became smaller, more reliable, and less expensive.

CONVENTIONAL CENTRAL OFFICE INTERFACES

All large systems are divided into modules or subsystems to simplify design, use and maintenance. The point where each subsystem connnects with another is an interface. A complete interface specification defines all mechanical, electrical, and operational rules for the inputs and outputs; ideally, it allows interconnection without requiring knowledge of the subsystem's internal operation. A large number of interfaces exist in the telephone network simply because of its size and complexity. And even more interfaces are required to adapt between old and new equipment and between the many different types of equipment that can accomplish the same function.

As we learned in Chapters 1 and 2, the central office is the place where all telephone calls are handled and first switched. If the call is to another subscriber in the same exchange, then the call is switched to that subscriber's line. If it is for a telephone in the same locality served by a different exchange, then the call must be switched to a trunk connecting the two central offices. If the call is destined for another city, it must be switched to the long-distance network via a toll-connecting trunk as illustrated in *Figure 7-1.* As it performs its switching function, the central office has two important (and different) interfaces. These interfaces are described best using telephone jargon. The local loop is commonly called the subscriber loop or the lineside interface; and the trunk to another switching office is called the trunkside interface.

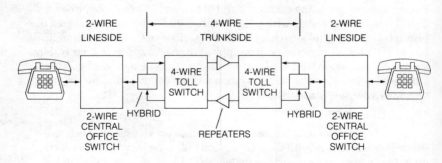

Figure 7-1. *Typical Analog Telephone Connection for a Long Distance Call*

LINESIDE INTERFACE

The largest number of interfaces in the telephone network occur between the telephone set and the local office. Because this interface has evolved through the days of magneto ringers, rotary dials, and step-by-step switches, the lineside interface has been more difficult to replace with electronics and still meet the standards and characteristics that have evolved over the years. Because there are so many local loops, this is the interface that will be part of the network longer than any other.

The basic requirements of the lineside interface in the conventional telephone network are referred to as BORSCHT, which we learned in Chapter 6 means Battery, Overvoltage protection, Ringing, Supervision, Coding, Hybrid, and Test. Let's now examine these functions of the conventional lineside or subscriber interface in more detail.

Battery Feed

The battery feed must provide the following for the local loop:

1. Power (typically 48 Vdc) to the subscriber's telephone set.
2. The capability to allow signaling to and from the telephone set.
3. Low dc resistance.
4. High ac impedance.

Various battery feed arrangements were discussed in Chapter 3.

Overvoltage Protection

Overvoltage protection protects equipment and personnel from dangerous transient voltages due to lightning surges of up to 1,000 volts and induced voltages from, or short circuits to, utility electrical power lines. Except for higher power ratings, the protection devices for the central office are similar to those described in Chapter 3 for the telephone set.

Ringing

In Chapter 2, the telephone set ringer and ringing generators in the central office were discussed. The central office must provide the ringing signal to the subscriber telephone to alert the called telephone that a call is waiting; therefore, the central office must apply the ringing signal to the line after the switching has completed the connection. This is done normally by a relay that is energized by the switch. The ringing signal is typically 90 volts rms at 20 hertz.

Supervision

Detecting service requests (when the caller goes off-hook), the dialing input, and supervising calls in progress (when the ring is answered, or when either party hangs up), is accomplished by detecting the presence or absence of current flow in the loop. This requires a sensor that can discriminate accurately, regardless of line length, between off-hook current and current as a result of noise, leakage or a small standby current for the memory in an electronic telephone. There are two common methods used for detecting a subscriber off-hook; loop start and ground start.

Loop Start

Loop start lines are used in the vast majority of local loop circuits; they signal off-hook by completing a circuit at the telephone. *Figure 7-2* illustrates a subscriber line interface using relays for sensing and logic. In the on-hook condition, neither the line relay nor the cut-off relay is operated, and the line relay battery provides power to the line. No current (except perhaps leakage current) flows because the switchhook contacts are open. When the subscriber lifts the handset *(Figure 7-2a)*, current flows from the line battery through the closed switchhook contacts and energizes the line relay. A set of the line relay contacts close to signal the switching circuits, via the line finder, that the subscriber wishes service. When the line finder seizes the line to

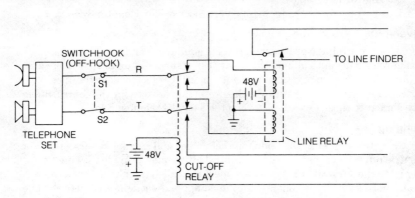

a. Line Relay Energized to Signal Line Finder

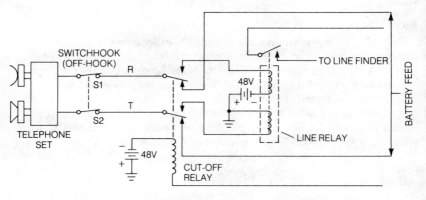

b. Cut-off Relay Energized to Extend Circuit into Switching Equipment

Figure 7-2. *Line Interface Relay Operation*

provide dial tone, it causes the cut-off relay to operate, which disconnects the line relay as shown in *Figure 7-2b*, and extends the circuit into the switching equipment. This also disconnects the line battery so further operation is powered by another battery supply through the first selector or register of the switching system. Note that the line relay has two windings, one connected in each side of the line. The windings are balanced and wound so that voltages induced in the line are canceled; thus, only current that flows around the entire loop will cause the relay to operate.

Ground Start

Ground-start lines are used on loops connecting PBXs to the central office, and in other situations where it is desirable to detect a line that has been selected for use (seizure of the line) instantaneously from either end of the line. Grounding the ring-side path, as shown in *Figure 7-3*, causes current to flow through one-half of the line relay which is sufficient to energize the relay. Further operation is as explained for loop start. When dial tone is detected by the PBX equipment, the ground-start contact is opened.

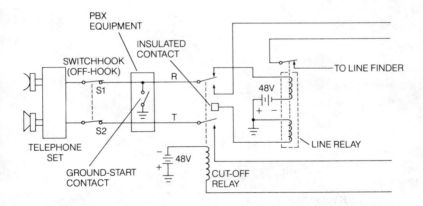

Figure 7-3. *Ground-Start Signaling*

Dialing Supervision

The problem in dial pulse signaling is to detect the difference between no current due to dial pulse break intervals and no current due to on-hook. This has been accomplished in the conventional system through the use of three relays, called A, B, and C, as shown in *Figure 7-4*. The A relay is fast-operating; it energizes when the tip-ring circuit is closed by the switchhook contact, then releases or energizes as the dial pulsing contacts open and close the circuit. Thus, the A relay "follows" the dial pulses and its contacts open and close in synchronization with the dial pulses to control the first selector in a step-by-step switch.

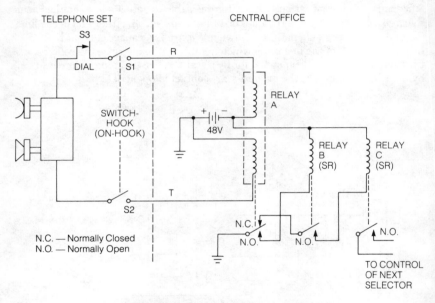

Figure 7-4. Dialing Supervision

The B relay energizes when A energizes, but it is a slow-release relay. It is designed so that enough energy is stored to hold its contacts closed for a short time after energizing current is removed; therefore, its contacts remain closed while the A relay contacts are opening and closing. The B relay does not release until the A relay has been released for about 200 milliseconds; thus, the B relay is held energized as long as the subscriber is off-hook and indicates, by releasing, when a subscriber hangs up.

The C relay also is a slow-release type and is used to detect the end of the pulse train for the dial-pulse receiver. When the A and B relays are already energized, the C relay is energized when the first dial pulse interrupts the current to release the A relay. The C relay remains energized as dial pulses are generated and does not release until the A relay remains energized for an interval of about 200 milliseconds. Since the longest standard dial pulse period is less than 100 milliseconds, a 200 ms release time for the C relay is sufficient to hold the C relay energized over the longest dial pulse interval. Then, when dialing is finished and the A relay remains energized, the C relay releases which signals the switch that dialing of that digit is completed.

Answer Supervision

Answer supervision involves disconnecting the ringing current (called "ring trip") when the called party answers; it also may require momentary interruption of the circuit, or reversing the polarity of the tip-ring pair. Some telephone companies use this means to indicate that the call has been "cut through" to the called party, so that charging for a toll call can begin. The possibility of polarity reversal means that the purchaser of a telephone set should insist that it be equipped with polarity protection as discussed in Chapter 3; otherwise, the set may not be able to signal remote computers or specialized carriers through the DTMF keypad. Worse yet, the set may be damaged if connected to a line where polarity reversal is used.

Coding

The coding of the voice signal into serial digital codes that are placed into PCM time slots for digital transmission occurs in an encoder at the sending end as discussed in Chapter 6. Recovering the signal at the receiving end requires a decoder. When both encoder and decoder are combined into one integrated circuit, it is called a codec. Codecs will be discussed in more detail later in this chapter.

Hybrid

The conventional local office is a two-wire switch, meaning that conversation travels in both directions over the same pair of wires. When a signal is to be transmitted over long distances, amplification is required; therefore, switches for these long-distance circuits must be four-wire as shown in *Figure 7-1*. The two-to-four-wire conversion is accomplished by the hybrid transformer as described in Chapter 2. The hybrid is on the trunkside interface of the conventional switch because local calls do not require conversion; only calls needing amplification because of distance must go through a hybrid. However, if the switch is digital, the hybrid must be part of the lineside interface.

Test

Testing requires access to the local loop circuit and to the circuits of the switching equipment to detect faults and provide maintenance. Additional relays placed in the local loop circuit provide the access required.

TRUNKSIDE INTERFACE

The word "trunk" has acquired many meanings (some conflicting) over the years. The most common use means a channel between the equipment at two switching locations. This definition still creates confusion, since for example, a "trunk" at a private branch exchange (PBX) installed in an industrial plant or in a business office ends up as a "line" at the central office, while a private circuit or "tie-line" between two PBXs is functionally identical to a trunk between two central office switches.

Trunk Circuits

Usually a trunk is terminated in a "trunk circuit", which is the interface for transmission, supervision and signaling between the trunk and the switching system. It is considered to be part of the trunk when measuring transmission levels. Trunks are more expensive than subscriber loops; thus, they are provided only where needed. However, because the trunk circuit is not dedicated to one customer, they usually have much higher usage than local loops. For example, a group of 30 trunks which are designed so that only 3% of the calls during peak traffic will be blocked (not connected) will average nearly 70% usage over a 24-hour period.

Short distance trunks may be pairs of wires, while long-distance trunks are usually implemented through multiplexed analog or digital carrier systems. The trunk itself may be one-way or two-way, and either automatic or operator-handled. The trunkside interface at the central office accommodates these varying types, and provides the same sort of functions that the lineside interface provides, although with more variations and complexity.

Figure 7-5 illustrates some of the interfaces on the lineside and trunkside of AT&T's computer-controlled space-division No. 1 Electronic Switch System (No. 1 ESS) showing typical switching paths and some of the terminology. Connections are routed through crossbar switches to connect line-to-line or line-to-trunk. Battery, supervision, signaling, and termination all are required on trunks.

ELECTRONICS IN CENTRAL OFFICE INTERFACES

For a number of years, the No. 1 and No. 1A ESS have been the principal class 5 switches in the Bell System. New designs have produced the No. 10A RSS (Remote Switching System). It is designed to be controlled remotely using a No. 1 ESS as the main controller. The No. 10A provides essentially all central office functions, and allows the use of electronic switching in areas where it formerly was not economical to do so. The No. 10A is, in fact, another example of distributed intelligence in the telephone network.

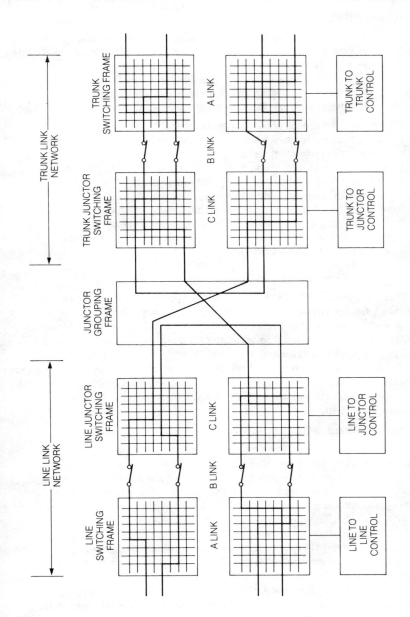

Figure 7-5. *Interfaces of Computer Controlled Switch*

Electronic SLIC

Our interest in the No. 10A is its subscriber line interface circuit (SLIC) and switching electronics. The switching portion will be discussed later in the chapter, but for now let's use the No. 10A's line interface, shown in *Figure 7-6*, to illustrate various ways of providing SLIC functions. The No. 10A SLIC is a circuit that is intermediate between a full integrated circuit and no electronics because it combines electronics with a transformer to provide:

1. Low power dissipation of 650 mW per line.
2. Isolation from common-mode signals on the local loop by transformer coupling to the power converter.
3. Protection against high voltage transients by diodes and transformers because of a transformer's low frequency response.
4. Two modes of operation; high power and low power. In high power operation, the power converter operating frequency changes from 40 Hz to 90 Hz depending on the line resistance. This varies the output voltage as a function of line length. In the low power mode, the converter only supplies enough current to compensate for loop leakage.
5. Application of ringing voltage.
6. Access to the circuit for testing.

During ringing and loop testing, the power feed from the converter and the voice feed must be disconnected from the loop. This is accomplished by stopping the converter and allowing the disconnect Triac to open. A ringing voltage of 20 Hz is applied from a common ringing bus via reed relays. Reed relays also provide test access. These relays are controlled by the remote switching system via latches and relay drivers on a control and timing IC. This control and timing IC also controls the converter IC, participates in setting up the path through the switch, and multiplexes the supervisory signals, including ring trip.

Integrated Circuit SLIC

The design of integrated circuits to satisfy the subscriber line interface functions has been directed not only to satisfy as many of the BORSCHT functions as possible, but also to look ahead to the conversion of more and more of the telephone system to digital. In addition, the flexibility of programmable digital systems to be changed to different applications or incorporate additional performance features by changing the program, has influenced the design.

However, there are limitations. Semiconductor materials, especially in the structure of integrated circuits like the SLIC that handle a variety of functions of amplification and logic, are limited in the breakdown voltages that they can withstand. Therefore, the high voltage and voltage transient protection normally are not provided on the integrated circuit, but are taken care of by external components. Integrated circuit differential amplifiers have very good common-mode rejection, but have difficulty handling the large common-mode signals appearing on the local loop. Therefore, the transformer

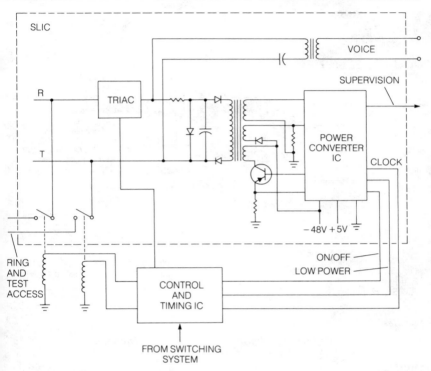

Figure 7-6. *Bell No. 10A RSS SLIC*

isolation available on the incoming local loop is maintained when using many integrated circuit SLICs and the battery feed available at the central office is used as is, rather than using the power converter technique described above. When SLICs provide battery feed, the most common technique is to provide current drive from external transistors to make sure adequate current is supplied the subscriber's telephone set for proper operation.

The TCM4204

An example of an integrated circuit designed to provide the subscriber line interface functions is the Texas Instruments TCM4204. As shown in *Figure 7-7*, the TCM4204 is designed to interface a subscriber line to a digital PCM switch and is intended to interface to a codec. It provides the hybrid, supervision and control of the ring and test functions. Because it provides the hybrid 4-wire to 2-wire conversion, this function moves from the trunkside interface to the lineside interface of the central office switch and is provided for each subscriber line. Battery feed stays as it exists in the central office and transient voltage protection is provided by an external TCM4301 integrated circuit.

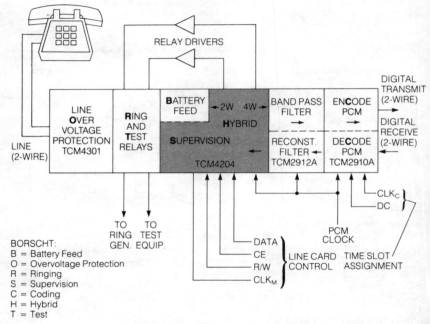

Figure 7-7. Digital Line Card Block Diagram

Operation

A more detailed diagram of the integrated circuit itself interfaced to the subscriber line is shown in *Figure 7-8*. Speech signals pass to the subscriber from the central office switch through the receive input (REC IN), through the receive path attenuator (REC ATTN), and through a driver that drives the line transformer in push-pull through external resistors R1 and R2. The external Zener diodes, D1 and D2, provide overvoltage protection. After transformer coupling in the central office, speech signals from the subscriber are input to the TCM4204 input op-amp, A1, through the transmit attenuator (XMIT ATTN), and out on the 2-wire transmit line through the switch to the trunk circuit. The gain of the operational amplifier is set by external resistors R3 and R4.

Switchhook closure (off-hook) is detected by the resistor bridge (R8 through R13) in the local loop side of the line transformer and fed through the operational amplifier, A2, to a low pass filter and to the microprocessor control bus through the digital interface. The low pass filter blocks the ringing signal when it is present so it will not affect the microprocessor. Dial tone is provided to the line through the REC IN input and then coupled through the receive path. DTMF dialing tones follow the transmit path and are output on XMIT OUT.

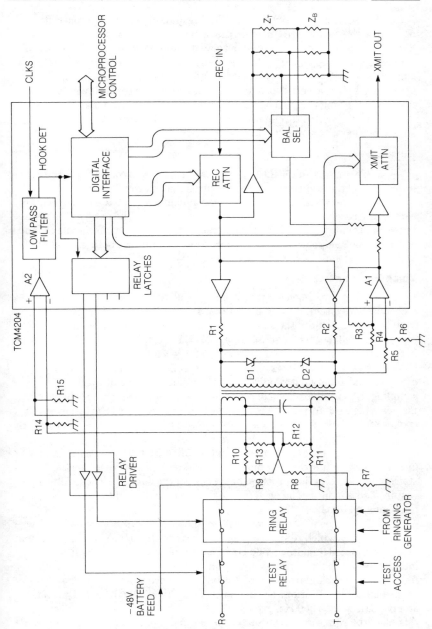

Figure 7-8. *Subscriber Line Control Circuit*

External relay drivers for the ring and test relays are controlled by on-chip latches which are timed by the microprocessor control through the digital interface. When activated, a relay driver energizes the ring relay to connect the ringing signal to the line. When the subscriber answers, off-hook current is detected by external resistor R7 to provide a ring trip signal through A2. Test access is provided by energizing the test relay in the same fashion as the ring relay. When the relays are not energized, battery feed of − 48 volts is provided through the normally closed relay contacts to the subscriber loop.

Advantages

The most significant feature of such a line control circuit is the digital interface and the additional control features that it provides. It allows the external microprocessor (which is shared among many SLICs) to separately control the receive and transmit attenuators, to select the external balance network providing Z_T and Z_B, to power-up or power-down on command, and to control analog loopback for testing purposes. In addition, both ground-start and loop-start supervision are provided.

Voice-Frequency Filters

When sending a conversation to a called party in any sampled transmission system, it is important that no frequencies higher than half of the sample rate be input to the sampling circuit in the encoder. If input, these frequencies will not be reproduced properly, but a spurious signal will appear in the output that is the difference between the actual unwanted frequency and half of the sample rate. This phenomenon is called *foldover distortion* or *aliasing*, and avoiding it requires that the input signal be filtered before sampling. (A visual example of the phenomenon of aliasing occurs in western movies where the frame rate is too slow to properly record the fast-moving spokes of the stagecoach wheels; thus, instead of rotating rapidly forward, the wheels appear to rotate slowly backward or even stop.)

When the signal arrives from the called party, the output of the decoder must be filtered to remove the high frequencies caused by the stairstep reconstruction process that occurs in the decoder. These filtering requirements indicate the need for inexpensive, high-quality filters in digital telephone systems.

The transmit and receive filters shown in *Figure 7-7* are basically low-pass filters which reject frequencies above 3 kHz. However, the transmit filter must also attenuate any 60 Hz input component (which might be induced into the local loop from nearby power lines); thus, it is actually a bandpass filter with a passband from 300 to 3,000 Hz. As with most filters, the performance specifications are in terms of passband (300 to 3,000 Hz) and stopband (less than 300 Hz, greater than 3,000 Hz) attenuation. However, in this case, the ripple specification (variations in response across the passband) is also very important because of problems with transmission quality when call routing results in several filters being placed in series, which is likely to happen in PCM transmission systems. The stopband requirements prevent aliasing of frequencies above 4 kHz.

The filters also contribute to crosstalk and idle channel noise. If more than one filter is put in a single monolithic device, care must be exercised to provide a large crosstalk attenuation between them to ensure that they do not interact. Crosstalk coupling through a common power supply also must be avoided.

The TCM2912A

A block diagram of a typical single-chip voice-band filter, the Texas Instruments TCM2912A, is shown in *Figure 7-9*. The diagram is divided into three sections. The transmit section has third-order high-pass and sixth-order low-pass filters to provide bandpass filtering to eliminate unwanted switching and low-frequency noise. The receive section furnishes sinx/x correction for the codec and eliminates high-frequency switching signals. The TCM2912A is designed to implement the transmit and receive passband filters for PCM trunks or line terminations. It uses switched capacitor techniques and is fabricated using NMOS technology. Transmit gain can be adjusted by the ratio of resistors R1 and R2. If high impedance electronic hybrids are used, the receive output (HIGH IMPEDANCE REC ANLG OUT) can drive the hybrid directly. If low impedance coupling is required, the on-chip power amplifier can be connected as shown in *Figure 7-9*. R3 and R4 are adjusted for required gain and R5 is for impedance matching. The third section of the diagram contains clock generators, voltage regulators for receive and transmit filters, and a substrate decoupler to reduce crosstalk.

Codecs and Combination Circuits

Figure 7-7 showed the block diagram of a digital line card. It showed the encoder and decoder combined into the codec (TCM2910A) and the codec interfacing with the TCM2912A. The TCM2912A contains both the transmit path filter and the receive path filter. The subscriber line control circuit, TCM4204, provided the required 4-wire to 2-wire conversion.

The arrangement of *Figure 7-7* is on a per-line or per-channel basis. There is also an arrangement whereby one codec is shared by a number of lines by using multiplexing techniques. This saves on equipment costs because fewer codecs are required; however, some problems exist when using shared codecs. The first is reliability or downtime. If many lines are multiplexed through one codec, many lines go down if one codec goes down. The second problem is that the analog time division multiplexing is much more difficult and less flexible than digital multiplexing. Also, shared codecs are more difficult to design in single-chip integrated circuit form due to the problems of crosstalk between the channels and obtaining the speed performance required.

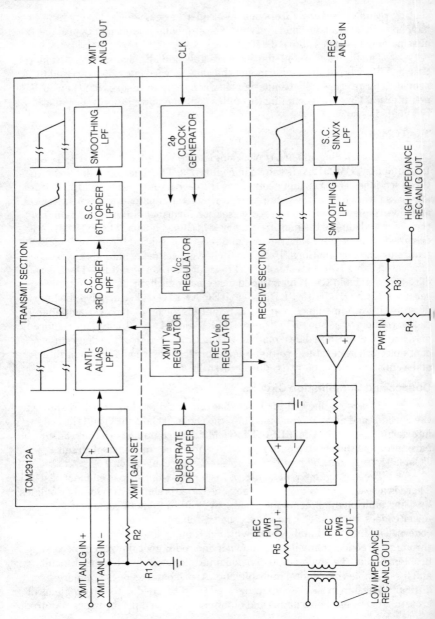

Figure 7-9. PCM Line Filter

Because of these problems and the potential volume (there are some 600 million subscriber telephone lines) which should reduce cost, most of the development has been directed to individual codecs per line. (The projected cost reduction follows the standard learning curve design philosophy where cost decreases as volume increases.)

The main applications of the codec in telephone systems are in central offices, channel banks, private automatic branch exchanges (PABX) and digital telephones. There may be some other non-telephone applications, but they are low-volume usage.

Operation

Virtually all codecs use the successive approximation technique illustrated in *Figure 7-10*. In the transmit direction, the two-wire signal from the subscriber is sampled in the sample-and-hold amplifier and encoded using the voltage comparator, the companding digital-to-analog converter, and the successive approximation register. The resulting compressed binary data are loaded into the data buffer and shifted out into the proper PCM timeslot under

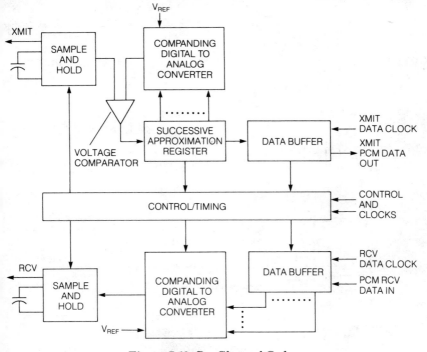

Figure 7-10. *Per-Channel Codec*
(From Paul R. Gray and David G. Messerschmitt, "Integrated Circuits for Local Digital Switching Line Interfaces", IEEE Communications, Vol. 18, No. 3, 1980, pp. 12-23, Copyright© 1980 IEEE)

control of the transmit data clock. In the receive mode, the digital information comes from the line to the expanding DAC via another buffer, and is timed into the buffer by the receive data clock. The companding DAC, acting like a A/D converter, produces a voltage which is held in the receive sample-and-hold amplifier, and then is routed through the reconstruction low-pass filter for smoothing.

The companding coding was discussed in Chapter 6 and *Figure 7-11* adds some more detail. The positive signal μ-law curve of normalized input versus output is used as an example because it is the most common in the U.S. The curve is implemented in the codec by a segmented linear approximation.

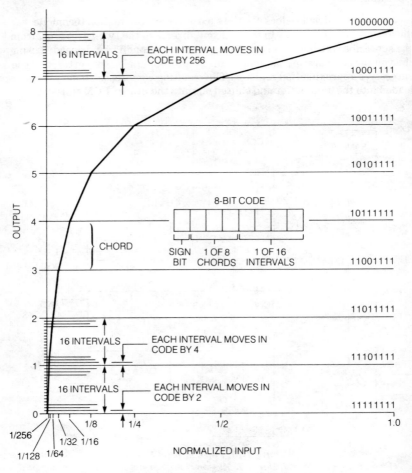

Figure 7-11. μ-*law Companding Code Detail*

7

ELECTRONICS IN THE
CENTRAL OFFICE

The total output signal range is divided into 16 segments called chords; 8 for positive signals and 8 for negative signals. Each chord is divided into 16 intervals. Each interval in a particular chord moves in output code by the same number of bits, but as the output increases in magnitude and spans a higher number chord, the coding produces a larger number per interval. As shown for the second output segment (chord 2), the output code increases by 4 for every interval, while for the eighth output segment (chord 8), the output code jumps 256 for every interval. The companding code internal to the codec is made up of 8 bits as shown in *Figure 7-11*. One bit is for the sign (+ or −), three bits are to identify the chord, and four bits are to identify the interval.

Combined Codec and Filter IC

Figure 7-12 shows the block diagram of a combo. It is called a combo because it combines the codec and filter functions into one IC. The filter functions can be recognized quickly by comparing *Figure 7-12* to *Figure 7-9*, and the related functions of *Figure 7-10* can be identified easily for the codec of *Figure 7-12*.

The TCM2914 is compatible with the D-type channel banks and asynchronous clocks used by AT&T. The TCM2913 is designed for use with synchronous clocks. Either μ-law or A-law companding can be selected with the μ-law/A-law select pin. Unless this pin is tied to the V_{BB} supply to select A-law, the companding is by μ-law.

In Chapter 6, *Figure 6-7* showed how signaling occurred in the bit stream. It showed the 8th bit of the code word being used for signaling and that this bit signaling occurred every 6th frame. This is called A signaling. There is also a B signaling per channel giving the possiblity of four different conditions being identified by the A and B signaling per channel. B signaling is also the 8th bit, but it is sent in the 12th frame. Thus, signaling is sent every 6th frame and the framing bit is used to identify the 6th, 12th, 18th, 24th frame, etc. Virtually all μ-law codecs provide for the insertion of the A and B signaling bits. On the TCM2914, the insertion for transmit is made on the μ-law/A-law pin when μ-law companding is selected and the signal output for the receive channel is on the Signaling Bit Out pin in *Figure 7-12*.

DTMF Receivers

Chapter 4 discussed the circuits in the telephone set that generate the DTMF dialing tones and showed that the DTMF generators were relatively easy to design and understand. Unfortunately, the reception and decoding of the tones at the central office is not as easy and this tends to make the DTMF receivers expensive. However, this is not a serious drawback to the use of DTMF, because each DTMF receiver is shared by approximately thirty subscriber lines. Also, recall that DTMF tones are used to transmit low-speed data to data communications equipment; therefore, DTMF receivers are required for these applications in addition to dialing.

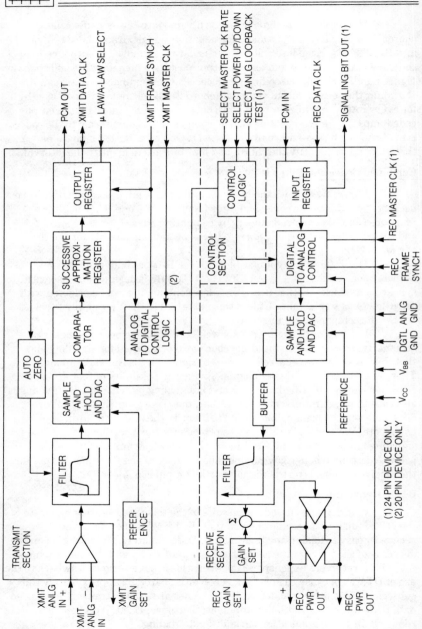

Figure 7-12. *Combined Single-Chip PCM Codec and Filter (TCM2913 and TCM2914)*

As DTMF tones are used more and more for data communications, the demand for DTMF receivers will increase. With the increased manufacturing volume, their cost should decrease.

Figure 7-13a again shows the DTMF frequencies generated at the telephone set. The frequencies in each horizontal row (low-frequency group) and each vertical column (high-frequency group) are separated by intervals of approximately 10%. The low-frequency and high-frequency groups are separated by about 25%. These specific frequencies have been selected with a great deal of care to meet several criteria, but one of the more important requirements is to have the minimum amount of harmonic interaction.

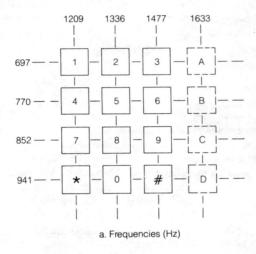

a. Frequencies (Hz)

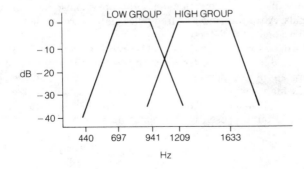

b. Idealized DTMF Receiver Filter Characteristics
(Reprinted with permission of *telephony* magazine)

Figure 7-13. DTMF Frequencies and Filter

A DTMF receiver must do the following:

1. Detect the tone-pair signal properly if the frequencies are within ±2% of the nominal values, and reject the signal if the frequencies are outside the limits of ±3%.
2. Make sure that one and only one tone is present from each group, and that the tone duration is at least 40 milliseconds (ms).
3. Detect as two separate signals any separation of valid tone pairs which is 35 ms or greater. Detect a tone pair signal as the same signal, not as two distinct signals, if the separation between the tone pairs is 5ms or less.
4. Properly detect tones whose level may vary over a 27.5 dB dynamic range. If the two fundamental frequencies of the DTMF pair have a difference in amplitude, it is called twist. The DTMF circuit must detect the tone pair with up to 6 dB of twist.
5. Properly detect DTMF signals in the presence of speech and noise.

Filtering and Detection

It should be obvious from the above requirements that filtering is an important function of the DTMF receiver. *Figure 7-13b* illustrates the frequency response of an ideal DTMF filter. Such a device is called a bandsplit filter because its output is in two separate bands of frequencies to pass the high and low groups of DTMF tones while rejecting other frequencies. The output of the filter must go to a detector to accomplish the other requirements.

Figure 7-14 is a generalized block diagram of a bandsplit filter feeding a detector in order to accomplish the two required functions. Semiconductor manufacturers are producing integrated circuits to provide both the filter and the detector functions. (Examples are the Mitel MT8865 and American Microsystems, Inc. AMI S3525 bandsplit filters, and the Mitel MT8860 decoder.) The 3rd and 5th order filters in the IC are similar to those used in the digital line card in the encode and decode path. It provides signal gain, input filtering for noise and speech rejection, and separate (bandsplit) filtering for the group signals. After filtering, the frequency signals are formed into square waves before passing to the detector.

At the detector, each group signal is processed separately. Digital techniques are used to determine frequency by counting the number of master clock pulses present in each period of the unknown frequency. The detected frequency of the high group and that of the low group are logically tested to determine if they are valid DTMF frequencies. If so, they are combined to produce a coded output representing the dialed number.

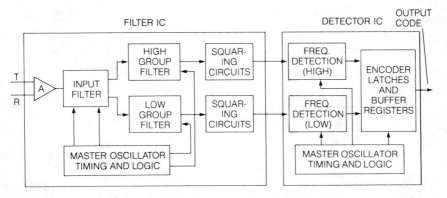

Figure 7-14. *DTMF Detection*

One of the more difficult tasks of the detector is to determine if a tone within the DTMF bands is really a DTMF tone or is a sound produced by speech that merely resembles a DTMF tone. If the "accept" criteria are too relaxed, DTMF tone-pairs are quickly recognized as valid, but so are speech segments or other sounds which "sound like" a real DTMF signal. Such detection is referred to as "talk-off", and can lead to wrong numbers or other erroneous operation at the receiver. If the "accept" criteria are too strict, detection time is stretched and the receiver no longer meets timing requirements. The result is a compromise between the effects of noise overriding true signals and the chance of talk-off errors. The detection time is typically about 10 milliseconds.

ELECTRONIC CROSSPOINT SWITCHING

In Chapter 1, the crosspoint switching was shown to be supplied by electromechanical crossbar switches and by reed relay switches. Reed relays are still being used by some telephone companies in their new equipment; however, most manufacturers have adopted some form of *true* electronic switching. An example of such a device is shown in *Figure 7-15*.

The basic switching element is a PNPN semiconductor device that is represented schematically as shown in *Figure 7-15a*. Each PNPN device is equivalent to two transistors, a PNP and an NPN, connected as shown in *Figure 7-15b*. The device has the following characteristics: If a voltage is applied to the device—positive to the anode and negative to the cathode—the device maintains a high resistance and acts like a reverse-biased diode. It remains in this state until the voltage applied between anode and cathode is increased past the breakdown voltage or until a current is drawn from anode to gate. When a current flows from anode to gate, the resistance of the device between anode and cathode becomes very low and the device conducts like a forward-biased diode.

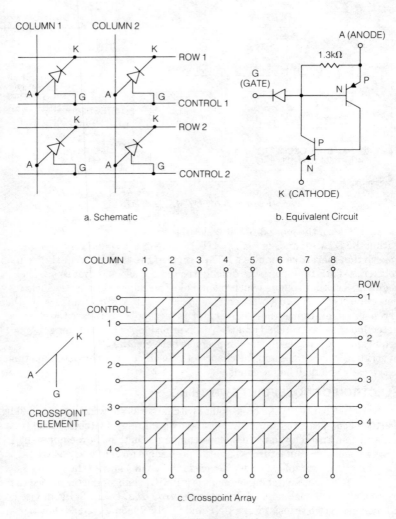

a. Schematic

b. Equivalent Circuit

c. Crosspoint Array

Figure 7-15. Crosspoint Elements and Array

Now, even if the current from anode to gate is turned off, the PNPN device maintains its low-resistance state until the voltage from anode to cathode is reduced to the point where a certain minimum current, called the holding current, cannot be maintained through the device. When the current through the device falls below the holding current value, the device changes back to its high-resistance state.

The device used in the No. 10A is manufactured in integrated circuit form[1]. As shown in *Figure 7-15c*, 32 crosspoints are fabricated in an 8-column by 4-row matrix. It is used as a low-resistance connection between the row and column leads by drawing a current out the gate lead while maintaining a minimum holding current from anode to cathode.

Figure 7-16 demonstrates how the arrays are used to connect 2-wire balanced speech transmission between two subscriber lines. Only partial arrays are shown and the detailed circuitry for the bottom half of *Figure 7-16* is just like the top half. The circuit connection is completed in two halves. Initially all points in the path are at a high voltage and all current sources I_h and I_t are off. The symbols for the current sources mean they supply a constant current of value I_h or I_t. I_h is the holding current and it is above the minimum value to hold the devices in a low-resistance state. I_t is the talking current that flows through the subscriber loop.

The connection is made as follows: First, I_h for the top section is enabled. Second, the logic gates, either G_A, G_B, or G_n turn on so that current can be drawn from gate to anode on the respective crosspoint elements. Note that any combination of crosspoints may be used to establish a path through the arrays. Third, the logic gate, G_1, controlling the driver for the center-tap of the input transformer turns on and the path is completed for the I_h current to be conducted to ground. This turns on and latches the PNPN devices, then the crosspoint logic gates are turned off. Now all crosspoints in the enabled path (shown in bold) are at low resistance and the top-half connection is completed. The same steps occur for the bottom-half circuitry of *Figure 7-16* to complete the switching that connects the calling subscriber line to the called subscriber line. Current sources I_t are switched on to support conversation on the lines. Speech is coupled from the top-half to the bottom-half by capacitors, but the dc is blocked.

The path connections are released by turning off the center-tap driver, disabling the current sources, and returning lines to a high voltage.

SUBSCRIBER LOOP SYSTEMS

One of the main applications of electronics in the subscriber loop is to extend service when the local loop becomes very long. As discussed in earlier chapters, all functions of local loop operation become marginal when loop resistance exceeds 1,300 ohms and loop current drops below 20 mA. In particular, DTMF dialing becomes unreliable, especially at low temperature, and transmitter gain suffers. The problems are most noticeable in the rural areas where line lengths of ten miles and longer occur frequently.

---

[1] J.M. Adrian, L. Freimanis, R.G. Sparber, "Peripheral Systems Architecture and Circuit Design", *The Bell System Technical Journal*, Vol. 61, No. 4, (April 1982), pp. 451-ff.

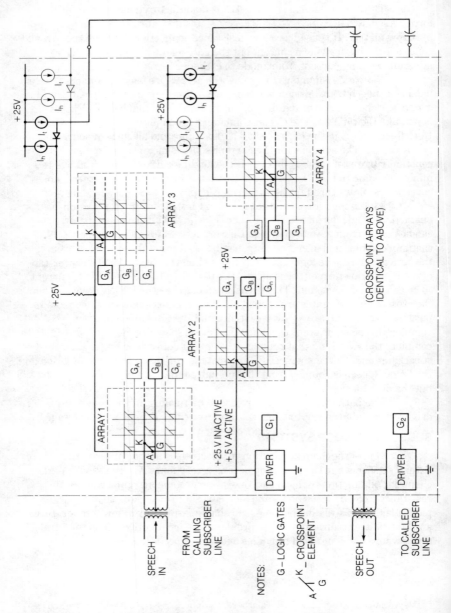

Figure 7-16. *Electronic Crosspoint Connection*

Electronics helped in the past by providing lower cost amplifiers to boost signal levels and by restoring voltage or current levels required for dialing and signaling. Newer techniques include multiplexing several (4 to 8) subscribers onto one local loop as discussed in Chapter 6.

More recently, microprocessors are being used in range-extender systems to provide automatic voice and signal gain at lower per channel cost. These systems automatically adjust themselves to provide the best transmission based on instantaneous measurements made at the time of call setup.

Digital Subscriber Loops

Digital carrier systems combined with local digital switching have eliminated the need for a multiplexer for the longer local loops and the conversion to an all-digital subscriber loop will aid this even further. In the previous discussions of digital transmission, the digital carrier brought the digital signals to the central office, where they interfaced with an analog subscriber loop. Digital subscriber loops allow the binary digits to come all the way to the subscriber's telephone set because all information is handled digitally. The telephone set itself contains the codec and filtering functions. Voice, data, supervision, and signaling are all in the same form and will be switched, transmitted and processed over the same distributed system. Such systems already exist in some PABX systems and eventually will be available for the public network.

WHAT HAVE WE LEARNED?

1. The subscriber line interface requires a wide variety of functions described as BORSCHT functions.
2. BORSCHT describes Battery feed, Overvoltage Protection, Ringing, Supervision, Coding, Hybrid and Test functions.
3. Integrated circuits are being designed to accomplish as many of the BORSCHT functions as possible on one chip.
4. Codecs are circuits which are being designed in integrated circuit form that do the digital encoding and decoding on the same chip.
5. Both high-pass and low-pass filters are required to reject noise and unwanted signals and to smooth the D/A output signal to reproduce the speech.
6. Detecting DTMF tones at the receiver is much more difficult than generating the tones at the telephone set because the receiver must be able to respond to real tones, but reject spurious signals that resemble tones.
7. A bandsplit filter has two separate passbands to separate the high-group and low-group frequencies of the DTMF tones.
8. PNPN semiconductor devices are being used to perform the crosspoint switching function instead of reed relays. The PNPN devices are being made as integrated circuit arrays.

Quiz for Chapter 7

1. The designer of central office equipment should:
   a. assume electronic telephones in metropolitan exchanges.
   b. change only one function of the interface at a time.
   c. maintain the standard interface between the subscriber and the CO.

2. The difference between the lineside and the trunkside interfaces is that:
   a. trunkside has the requirement of BORSCHT, and lineside does not.
   b. lineside interfaces are usually multiplexed analog carrier systems.
   c. lineside refers to the subscriber loop, and trunkside to connections to other switches.

3. The SLIC is:
   a. the speech loop interface card.
   b. the subscriber line interface circuit.
   c. the subscriber loop input command.

4. BORSCHT, as used in this book, stands for:
   a. battery, overvoltage, ringing, supervision, coding, hybrid, and test.
   b. a Russian beet soup fed to telephone company employees who work outdoors.
   c. battery, on-hook, ringing, subscriber interface, coding, hybrid, and trip

5. The principal difference between loop start and ground start is:
   a. loop start uses relays; ground start uses integrated circuits.
   b. loop start is one-way signaling which may involve a slight delay; ground start is either-way signaling that is instantaneous.
   c. loop start uses the line and cutoff relay; ground start uses the A, B, and C relays.

6. Foldover distortion or aliasing is:
   a. eliminated by filtering out frequencies below 300 Hz.
   b. another name for crosstalk.
   c. the presence of spurious frequencies caused by having too high frequencies in the sampled signal.

7. Codecs usually encode the signal by:
   a. sampling it and outputing the amplitude of the sample.
   b. folding the signal over and encoding it.
   c. sampling it and using successive binary approximations to encode.
   d. comparing analog waveforms to obtain the closest match.

8. DTMF receivers are:
   a. fortunately much simpler than DTMF generators.
   b. required to tell the difference between random speech frequencies and DTMF tones.
   c. designed to accept the digit if either of the two tones is present, to allow a margin for error in transmission.

9. Semiconductor crosspoint switching arrays:
   a. use PNPN devices that will remain latched in the conducting state.
   b. are used in the Bell No. 1A Electronic Switching System.
   c. are not yet practical, because of the high voltages that must be switched.
   d. are another name for subminiature reed relays.

10. Electronic crosspoint arrays require:
   a. a minimum holding current and a talking current to stay latched.
   b. a minimum holding current to stay latched.
   c. the difference between holding and talking current to operate.
   d. a voltage greater than 48V between anode and cathode to turn off.

Network Transmission

ABOUT THIS CHAPTER

This chapter describes the techniques, transmission link (media) and equipment used to transmit the digital signals described in Chapters 6 and 7 over long distances. It describes channel banks, multiplexers, and repeaters which are part of the electronic equipment used in digital transmission systems.

WHY DIGITAL TRANSMISSION?

In Chapter 6, a number of advantages and disadvantages of using digital systems were discussed. Despite the disadvantage of requiring eight times the bandwidth, the use of digital techniques is increasing rapidly in the network. Increased performance at lower cost using integrated circuits was pointed out as a prime advantage. This is having such an impact that shorter and shorter digital transmission lines between different types of offices are becoming cost effective. Lines as short as 10 miles are already economical, and if the switching equipment at each end is digital, digital transmission lines of any length will be practical. Less noise, lower signal to noise ratio requirements, and lower error rates are additional advantages.

Table 8-1. FDM Versus PCM Noise Comparison

	FDM/Radio/Cable	PCM/Radio/Cable
Multiplex	2,500 pWp*	130 pWp equivalent
Radio/cable	7,500 pWp	0 pWp
Total	10,000 pWp	130 pwp equivalent

*pWp = peak picowatts (10^{-12} watt) weighted according to the psophometric weighting curve

(From R.L. Freeman, Telecommunications Transmission Handbook, Second Edition, John Wiley & Sons, 1981, Copyright© 1981 by John Wiley & Sons, Inc., Reprinted by permission of John Wiley & Sons, Inc.)

Data from Freeman, listed in *Table 8-1*, indicates that the total noise generated in a 2,500 kilometer (km) standard frequency division multiplexed (FDM) system by the transmission media (cable and radio) is 10,000 pWp. The noise in an equivalent 2,500 km PCM transmission system is only 130 pWp – a reduction of almost 19 db. By adding repeaters in the line, the error rate is reduced. If the error rate is too high, spacing repeaters closer together will lower it. Thus, transmission errors become independent of total transmission line length.

A few additional advantages are worth noting:

1. Digital techniques contribute to the design of rugged transmission lines that are easy to design, easy to install and easy to maintain.
2. Digital transmission lines handle all varieties of source signals; broadcast music, data, and television, which will eliminate the need for special lines for special signals.

DIGITAL CHANNEL BANKS

Recall that in Chapter 6, channel banks were defined as the equipment at each end of a digital transmission that converts speech signals into coded bits. Equipment cost analysis has dictated the development of equipment that multiplexes many channels onto one wide bandwidth channel. For example, the standard system (T1 carrier) is made up of 24 individual voice channels. The channel banks generate and multiplex the digital signals. The Bell System channel banks are called D-type (for digital) and form the foundation for a whole series of multiplexers.

D-Type Channel Bank

The output from a D-type channel bank is a 1.544 Mbps digital signal called a DS-1 signal. The DS-1 signal might be transmitted on a T1 transmission line, which has repeaters in the line for continued signal regeneration, or it may be multiplexed with other DS-1 signals to produce a still higher level multiplexed signal. The DS-1 signal contains the bit stream with the digitized voice signals from each channel in their time slots and the channel signaling bits in positions as discussed in Chapter 6 (*Figure 6-6*)

D1 Channel Bank

Figure 8-1 is a block diagram of a section of a D1 channel bank. *Figure 8-1a* shows the transmitting portion, and *Figure 8-1b* shows the receiving portion. The transmitting portion performs three basic functions:

1. Digitally encoding the 24 analog channels
2. Adding the signaling information from each channel
3. Multiplexing the digital stream onto the transmission medium.

The channel inputs, gates, filters, compression circuits, and encoder provide filtering, sampling, compression (half of the companding function), and coding (analog to digital conversion). The signaling information from the channel signaling circuits, and the framing pulses from the master clock and framing generator, are inserted in the digital stream by the transmit converter. The resulting PCM stream is coupled to the transmission lines through a common control circuit. Since many different signaling schemes as well as both 2-wire and 4-wire trunks are in use, each of the five kinds of D channel banks; D1, D2, D3, D4, and DCT (Digital Carrier Trunk), contains interface circuits between individual voice channels and the common equipment shared by all circuits.

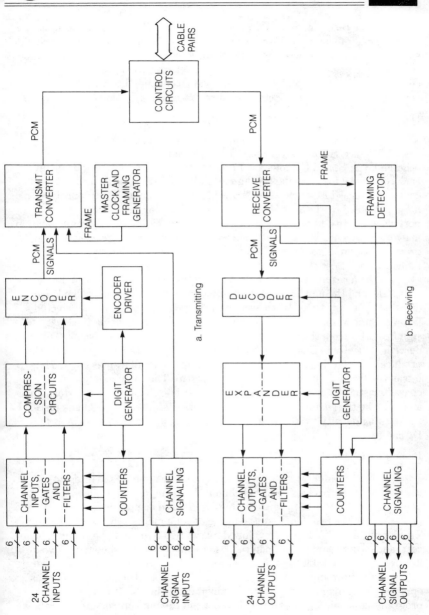

Figure 8-1. *D1 Channel Bank (24 Channels)*

The receiving portion reverses the process. It extracts the signaling and framing information, passes the signaling to the receive side of the channel signaling circuits, and passes the PCM encoded voice to a decoder (which does the digital to analog conversion), and expander (the other half of the companding function), and on to the individual channel filters and output circuits. A D1 channel bank contains three 24-channel sections (72 channels).

D Channel Bank Modifications

The success of the D1 banks, first installed in 1962, led to modifications (D1B and D1C versions), and then to a major revision. The original D1A, B, and C banks used 7 bits for each voice sample and one bit in each code word for carrying the signaling. When it became desirable to connect several T1 transmission spans together in tandem, the quantizing noise produced by the 7-bit encoding was too high for satisfactory performance. In addition, it was realized that providing signaling information in every code word is wasteful since 8,000 bits per second was not required to provide the signaling information for a channel because the signaling information did not change that rapidly.

As a result of these conditions, another modification to the D1 series (the D1D) and the new D2 channel bank were developed. The D2 bank uses all eight bits of every time slot to encode the analog signal except for selected frames. Supervisory and signaling information is sent by using the least significant bit from the code word in each channel every sixth frame. This was described briefly in Chapter 6, but is given in more detail in *Figure 8-2*. This produces an effective average code word length over all frames of 7-5/6 bits (8 bits in the first 5 frames plus 7 bits in the sixth). The companding characteristic also was changed to give better idle channel noise performance and to make the code words easier to linearize digitally. The D2 bank increased the packing density to 96 channels in the same space as the 72 channels for a D1 bank.

D3 and D4 Channel Banks

The D3 and D4 banks were motivated by advances in integrated circuits, allowing packaging of 144 channels in a single bay. Following the D4 bank, the application of programmable distributed computing and signal processing to the problem resulted in the development of the Digital Carrier Trunk unit, or DCT. It was developed by the Bell System to be smaller, lower cost, and easier to maintain than the D4 channel bank. As shown in *Figure 8-3*, it differs from the D4 unit in that it has no separate trunk circuits, no separate signal distributor and scanner, and it uses a single kind of channel unit—as opposed to the over 35 unique types available for the D4. It includes programmable microprocessors called peripheral unit controllers that relieve the main switching machine (usually a No. 1 or No. 1A ESS) of the task of sorting out the signaling information on the incoming trunks. The main switch

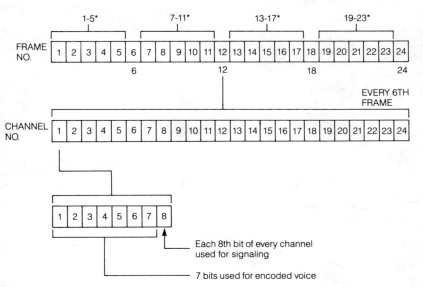

*All 8 bits of every channel in these frames are used for encoded voice.

Figure 8-2. *Sixth Frame Signaling of D1D Channel Bank*

also uses the peripheral unit controller to send the signaling, alarm and trunk status over a data link to a central control unit which routes the voice signals over the correct trunks.

MULTIPLEXERS

The electronic circuit that combines digital signals from several sources into a single stream is called a multiplexer. The basic circuit element called a serializer was shown in *Figure 6-25*. A multiplexer takes in many channels and allows each in turn access to the output to produce an output bit stream in serial sequence. Each input in turn is gated into the proper time slot in the output. If the time slot for a channel in the output is long enough to contain the entire input code word, the stream is said to be word interleaved. If the input is gated to the output one bit from each channel in sequence, the stream is bit interleaved. Since all timing is derived from the master clock, and all of the input streams run at the same rate, the unit is a synchronous multiplexer. All of the higher-order multiplex systems used in the telephone network are the synchronous type, but asynchronous multiplexers are used in data transmission applications.

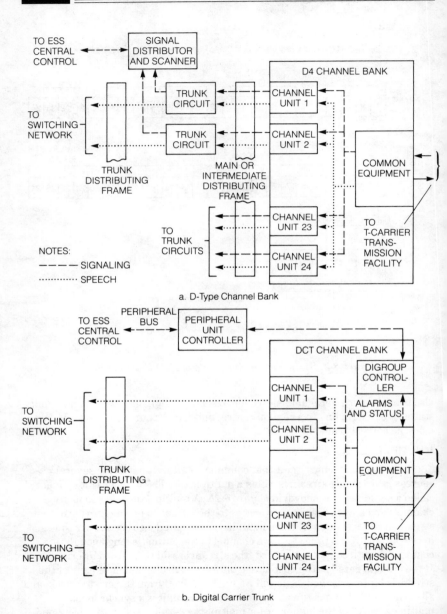

Figure 8-3. *Comparison of D-type Channel Bank and Digital Carrier Trunk*
(Courtesy of Bell Laboratories)

Framing Formats

Message channels are multiplexed into the bit stream in frame formats. Frames are set by the number of multiplexed channels that are packaged together. A D channel bank multiplexes 24 channels to form a DS-1 signal. CCITT multiplexes 30 channels into a frame. A special code, either a bit, word or words, is added in the frame to synchronize the frames of the PCM signal. Certain formats have been established for the time placement of the framing information. *Figure 8-4* shows three types—bit framing, bunched word framing and distributed word framing.

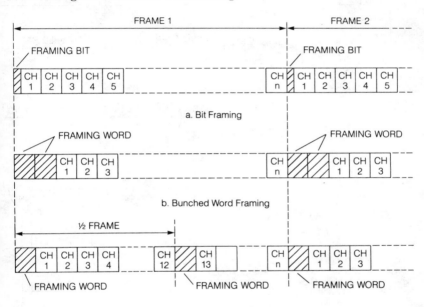

Figure 8-4. Framing Formats

Multiplexer Hierarchy

The multiplexers used in the public telephone network are arranged in an order that depends on the bit rate of the PCM signals that are transmitted. Multiplexer units with lower bit rates and fewer numbers of channels couple to units that have higher bit rates and greater number of channels. *Figure 8-5* illustrates the relationship of these systems. The output rates; and therefore, the numbers of channels that each multiplexer can combine, are matched to the data rates which can be carried on the various transmission media, from the DS-1 signal at 1.544 Mbps to the DS-4 at 274.176 Mbps.

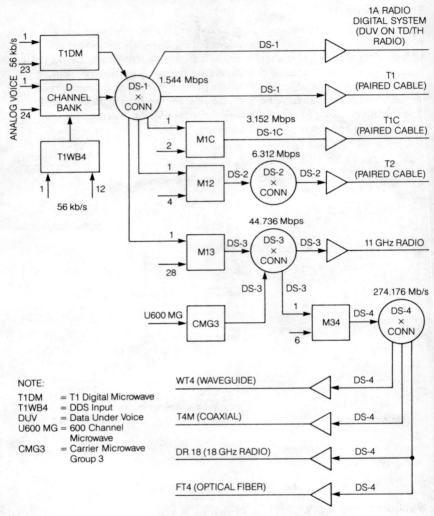

***Figure 8-5.** TDM-PCM Heirarchy*
(Courtesy of Bell Laboratories)

The multiplexing group members in the North American network above the D-type channel banks are:

1. The M1C which combines two DS-1 signals into a single DS-1C signal at 3.152 Mbps.
2. The M12 which combines four DS-1 signals into a single DS-2 signal at 6.312 Mbps.

3. The M13 which combines 28 DS-1 signals into a single DS-3 signal at 44.736 Mbps.
4. The M34 which combines six DS-3 signals into a single DS-4 signal at 274.176 Mbps.

Included in *Figure 8-5* are the different types of transmission lines used from T1 cable to coaxial, waveguide, radio, and optical fiber transmission media.

Pulse Stuffing

The timing for all operations within a multiplexer, including the clocking for the output bit stream, is provided by a master clock within the unit. The timing for the received data streams, however, is provided by each individual bit stream. The data rate for each of these input bit streams is likely to be slightly different from any of the other bit streams due to variations in the master clocks of the sources transmitting the signals and because of variations in the propagation delays of the circuits and media carrying the bit streams. Some method of allowing for these mismatches must be provided. In multiplexing, the method is called pulse stuffing.

The principle of pulse stuffing, illustrated in *Figure 8-6*, is to cause the outgoing bit rate of the multiplexer to be higher than the sum of all of the incoming bit rates plus any data added in the multiplexer for framing and synchronization. The incoming signals are stuffed with a number of pulses sufficient to raise their rates to be the same as the locally generated master clock. Extra bits (synch signal data channel), common to all channels, are provided to send the receiver the location of the stuffed pulses. The information about the location of the stuffed pulses is sent repeatedly (redundantly) to reduce the probability of an error that would cause loss of frame synchronization.

Each incoming bit stream is fed into a buffer (called an elastic store) at the incoming bit rate (Point A of *Figure 8-6b*) and temporarily stays in a stacked location. At appropriate times, a pulse is stuffed in the stream to provide the output shown at point B in *Figure 8-6b*. Point B's output is transmitted to the receiving unit and stored in another elastic store. A phase locked loop generates the clock which reads out the bit stream at point D. The time slots where pulses were stuffed are identified at the receiver by the stuff codes from the synch signal data channel and the pulses are unstuffed as shown in the waveform of *Figure 8-6d* for point C. The read clock rate of the receiver restores the bit stream to the uniform signal similar to that which was input at point A. If the buffer is ever empty, a pulse is provided so that the output can continue at a constant rate. A corresponding stuff code is sent on the synch channel to identify the stuffed timing bit. Certain formats and rules for detecting stuffing bits for each multiplexer type have been established. Let's look at the ones for the multiplexer for the DS-1C signals.

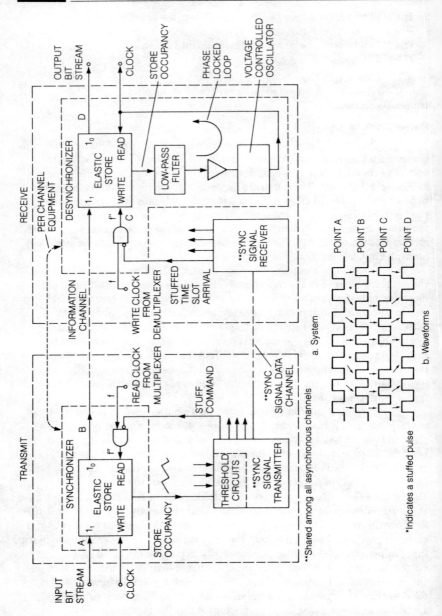

Figure 8-6. *Pulse-Stuffing Synchronization*
(Courtesy of Bell Laboratories)

DS-1C Multiplexed Signal

The master frame of the DS-1C PCM signal is made up of 1,272 bits divided into four 318-bit subframes as shown in *Figure 8-7*. M is the leading bit for the subframes and its pattern of 011X identifies the master frame and the subframe boundaries. The X bit is always a 1 unless an error occurs in transmission of the data, in which case the X bit is a 0. The regular bit pattern of 01010101 for F_0 and F_1 establishes the frames with the stuffing bits included for each subframe. The bits designated as stuffing bits are stuffed (interpreted as a null) when the bits shown as C bits are all 1 in a subframe. Otherwise, the stuffing bits are interpreted as data. It should be pointed out that the frame lengths for the DS-1C and higher signals are not related in any systematic way with the input DS-1 frames. The multiplexer treats the input DS-1 stream as simply a string of bits, passing data and synchronization pulses alike into the higher level stream.

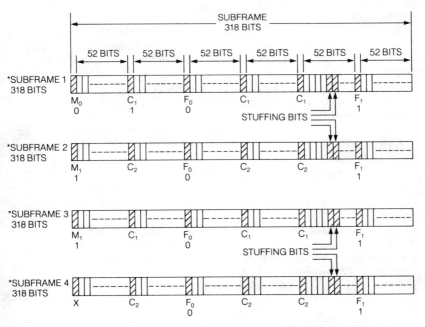

*Four subframes of 318 bits each = one master frame of 1,272 bits

***Figure 8-7.** DS-1C Frame Format Showing Pulse Stuffing Bits*

The higher level multiplexed signals have similar bit patterns established for master frames, subframes, frames and stuffing bit positions and stuffing rules. As with the DS-1C signal, the bits that form the pattern are distributed across the frame in a uniform way to minimize the probability of missing more than one stuff code due to a burst of errors. Such a miss would cause a frame slip which would cause a loss of synchronization.

Synchronization

The previous discussions have emphasized the importance of preventing loss of synchronization, or clock slips. The effect of these errors on digitized voice is not serious until the clock misalignments reach very large values (on the order of 1 part in 100,000). However, digitized signals from voiceband modems and voice or data signals which have been coded to maintain security (encrypted) are highly susceptible to clock slips. High-speed modem data are generally phase modulated, and a single 8-bit slip in the digital data produces a phase shift of 81 degrees. Not only are the data coded for security, but also the data are scrambled further because the modem loses synchronization and may take several seconds to recover. Such slips give audible blips in the voice signal and cause problems in the decoding equipment.

There are several possible methods for synchronizing a network of independent switching machines such as the telephone network. One method is to supply each machine with a clock so accurate that the difference between it and all other clocks is so small that slips seldom occur. This technique is called *plesiochronous* synchronization. It is expensive to implement since the clocks are expensive and must be redundant at each switch. Nevertheless, this method is the one chosen by the CCITT for international digital transmission in Europe and the gateways to the East. The clocks must be stable to within 1 part in 100,000,000,000.

Another method is to do pulse stuffing network-wide, as is done in the higher-level multiplex systems. This would require every channel at every digital switch to be stuffed separately on transmit. Such an approach is not economically feasible.

The method adopted for the North American network is that of master/slave synchronization. The master timing frequency reference is maintained in Hillsboro, Missouri, which is near the geographical center of the network. From there, the master timing reference is sent to selected switching centers, which become slave timing centers, over dedicated transmission facilities. From these slave centers, the timing reference is forwarded to lower level centers over existing digital facilities. *Figure 8-8* shows a network diagram for a master/slave synchronizing system for North America with the central reference designated as M. The numbers in the circles are the office classes of the network.

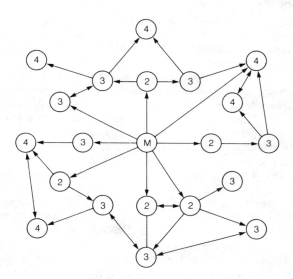

Figure 8-8. *Master/Slave Timing Network*
(From Luetchford, et al, "Synchronization of a Digital Network", IEEE Transactions
on Communications, *Vol. COM-28, Aug. 1980, Copyright© 1980 IEEE)*

LINE CODING

The waveform pattern of voltage or current used to represent the 1's
and 0's of a digital signal on a transmission link is called the line code. There
are many different types of line codes in use and several will be examined
in a moment, but first some basic requirements of a line code.

1. The level of direct current in the transmission medium must be constant,
   and preferably zero.
2. The energy spectrum of the wave must be shaped to avoid frequencies
   outside of the bandwidth available.
3. The signal must contain adequate timing information so that the clocking
   signal can be recovered by regenerative repeaters in the transmission path.
4. Errors should be easy to detect, and their detection used to indicate the
   performance (error rate) of the link.
5. Errors should not propagate; i.e., single-bit errors on the line should not
   cause multiple-bit errors in the decoder.
6. The coding scheme should use the code space efficiently; i.e., if possible, the
   scheme should use only two states to represent one and zero.
7. The coding scheme should minimize crosstalk between channels.

Types of Line Codes

Figure 8-9 details the different types of line codes. In reference to the bit stream shown in *Figure 8-9a*, a typical binary signal waveform is shown in *Figure 8-9b*. It is the simplest of the code types because it uses just two voltage levels; in this example +3V and 0V are used. It is called *unipolar* because it is not symmetrical about 0V. When the waveform is symmetrical about 0V as shown in *Figure 8-9c*, it is called a polar code. If the signal level representing each bit; e.g., +1.5V for a 1 and −1.5V for a 0, maintains the assigned value for the entire time duration allotted to the bit (time segment T), the code also is called a Non-Return-to-Zero (NRZ) code.

The code that is primary to the DS-1 multiplexer signal is the bipolar code shown in *Figure 8-9d*. It also is called Alternate Mark Inversion coding (AMI). It produces alternate positive and negative level pulses, symmetrical around 0V, when successive 1's occur in sequence. As a result, AMI is a three-level or ternary signal, where a 1 is represented by either a positive-going or a negative-going pulse in a signal interval, while a 0 is represented by the absence of a pulse in a signal interval. This scheme, first used by the Bell System in the U.S., satisfies most of the basic requirements listed previously. There is no dc component in the transmitted signal, the amount of energy in the signal at low frequencies is small, and compared with unipolar signaling, AMI has a substantial advantage in that it has much more immunity to crosstalk (on the order of 23 dB).

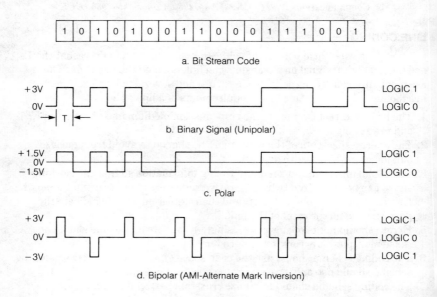

Figure 8-9. Types of Line Coding

The waveform of AMI pulses is shaped to occupy only about one-half of the allowable pulse width. This is done to simplify the circuits that recover the timing information in the regenerative repeaters. The total waveform does not have a 50% duty cycle as occurs in the clock waveform shown in *Figure 8-10a*, but the symmetrical pulses are said to have a 50% duty cycle.

DC Wander

A principal requirement of a line code is that it contain no dc, or if a dc component is present, it must be a constant level. This is because most carrier transmission systems do not pass dc. The problem that occurs when a large dc component is present is shown in *Figure 8-10*. Long sequences of 1's or 0's can cause a shift in the waveform amplitude with respect to zero volts. When this occurs, signal recovery circuits that are detecting the 1's and 0's lose proper amplitude reference and errors occur. Special circuits called dc restoration circuits are designed to eliminate the dc wander and restore the proper reference levels.

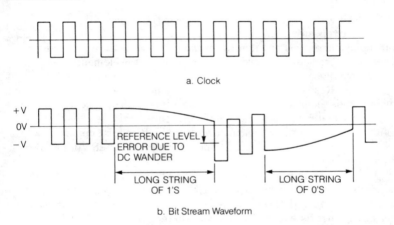

a. Clock

b. Bit Stream Waveform

Figure 8-10. *DC Wander*

PULSE AND TIMING RESTORATION

Digital signals have the useful property of always starting out with a well-defined and unique shape or waveform. The encoding scheme for transmission of the bits down a channel is designed so that only a small number (typically two or three) of such waveforms are possible. This makes it easy to distinguish genuine binary information from noise, even if the waveform is severely distorted. Signals sent down a cable are diminished in amplitude (attenuated) by an amount which depends in part on the frequency of the signal (roughly by an amount equal to 8.6 times the square root of the frequency). If the original signal has a fairly wide bandwidth, the attenuation will be greater for the higher frequencies than for the low.

Shifts in the time relationship of signals (phase) also occur as the signals are transmitted over a transmission link. Shifts in phase also contribute to distorting the signal waveforms. The process of compensating for the amplitude and phase shifts is called *equalization*.

In principle, the equalization compensation should have an amplitude and phase dependence on frequency which is the exact opposite to that which occurs as the signals are transmitted over the link. However, in practice, the equalization is quite good on amplitude but is not very good at phase correction.

Whenever the original waveforms are reconstructed, the signal is said to be regenerated. The process is not at all like mere amplificaton, although voltage amplitudes may in fact be increased. The significant fact is that the regenerated new pulses are just like the original ones in size and shape. They may be delayed in absolute time, but have the same position in time relative to each other.

Repeaters

The system components that perform the regeneration task are called regenerative repeaters. Effectively, they take an attenuated and distorted input signal and reconstruct it at the output to look like the original signal applied to the transmission link. *Figure 8-11* shows the elements of such a regenerative repeater. The two functions it performs are pulse reshaping and retiming. Reshaping restores the pulse shapes to their original form, and retiming sends them out at the proper time intervals. The repeater consists of impedance matching networks, amplifiers, threshold bias circuits, timing circuits, pulse regenerators, and a power supply. Power is supplied down the same pairs as carry the data signals, and is usually either -48 Vdc or -130 volts from the central office.

The incoming signal is amplified, and drives the detector threshold bias circuit, timing chain, and balanced pulse regenerator circuits. The threshold bias circuit determines the decision level for the incoming signal, which determines for each pulse interval whether or not a pulse will be produced.

Clock Recovery

The timing chain is illustrated in *Figure 8-12*. Its main purpose is to retime the PCM signal by recovering the clock based on the timing of the incoming signal. The equalized pulse train (balanced about plus and minus levels) is rectified, differentiated, clipped, and input to the resonant tuned circuit. The waveforms of the signals at the points A through G of *Figure 8-12* are shown in *Figure 8-13*. Point A is the input AMI signal, point B the rectified signal, point C the differentiated signal and point D the clipped signal. Only the positive going pulses remain to be used as trigger pulses to the tuned circuit.

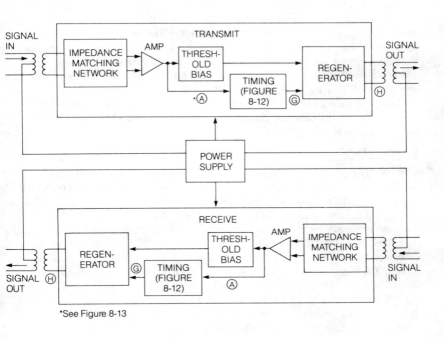

Figure 8-11. *PCM Regenerative Repeater*

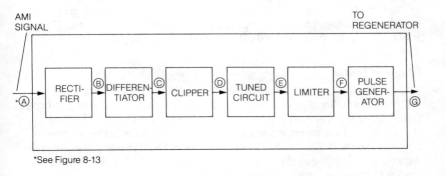

Figure 8-12. *Timing Circuit*

The input trigger at point D excites the tuned circuit and causes it to ring (oscillate) at its resonant frequency as shown in *Figure 8-13* waveform E. The ringing is at a constant frequency, but slowly decreasing amplitude. The constant frequency pulses are counted, divided, and sent through a limiter to produce the clock pulses of waveform F. Each time a trigger pulse arrives at the tuned circuit it restores the amplitude of the ringing signal. It is quite obvious that if a trigger pulse did not arrive for a long period of time, the amplitude of the ringing signal would decrease to zero. This would be the case if a long train of 0's occurred in the bit stream. The ringing amplitude would decrease below the detector threshold and the clock would be lost until the next 1 pulse arrived at the input.

Regeneration

The clock pulses of point F are input to a pulse generator which produces positive and negative pulses at the zero crossing points of the clock square wave. In telephone terms, such clock recovery is termed forward acting and the repeater is said to be self-timed. The clock pulses at the positive-going zero crossings of the timing wave are used to gate the incoming equalized pulses to the generator. The pulses from the negative-going crossings turn off the regenerator to control the width of the regenerated pulses. This action is called complete timing with pulse width control. The output at point H in *Figure 8-11* is a regenerated signal as shown in *Figure 8-13* that matches the original generated AMI signal.

As has been mentioned previously, it is possible to reduce the error rate on digital transmission facilities to any desired value by spacing the regenerative repeaters closer together. As a practical matter, however, adequate error performance for T1 spans is obtained by providing repeaters every 6,000 feet (1,828 meters). This particular distance is important because it was the spacing for the amplifiers for the previously used analog carrier system which T1 replaces. Thus, the same manholes can be used when a system is replaced. The energy in the DS-1 signal is concentrated at 772 kHz, which gives a loss figure of 26.6 dB at 6,000 feet over 22-gauge paired wire cables.

Received Signal

It is the job of the receiver at the end of the digital transmission to receive and amplify the transmitted signal and convert it to the original voice signal. In order to accomplish this task, the receiver must stay in frame alignment with the transmitted signal. In order to stay in frame alignment, it must extract its information from the transmitted signal and must operate with the same timing as the transmitter. It does this with a circuit very similar to the clock recovery timing circuit in the repeater.

A rectifier, differentiator and clipper form trigger pulses that feed a similar tuned circuit to produce a constant frequency timing signal. Square wave clock pulses are produced from the ringing signal with limiters and timing pulses are generated at the zero crossings of the square wave to produce accurate timing at the receiver. With accurate timing, the receiver is able to search and check to make sure it stays in frame alignment.

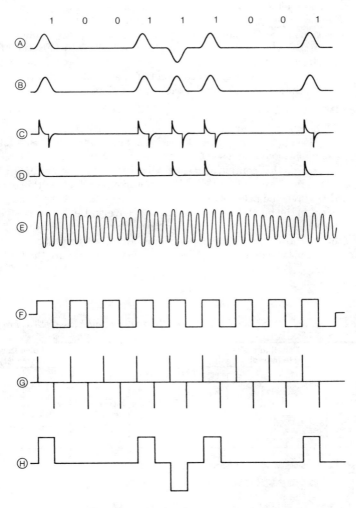

Figure 8-13. Timing Circuit Waveforms

Frame Alignment

Figure 8-14 shows a state diagram that describes how the receiver stays in frame alignment. As shown in *Figure 8-14b*, there is a word in the frame designated as a frame alignment word (FAW). If the receiver is out of frame alignment—state 5 of *Figure 8-14a*—the receiver begins a search for the FAW as shown in *Figure 8-14b*. If the FAW is detected in two successive frames, alignment is assumed and succeeding FAW positions are checked to verify alignment. This is the progression from state 5 to state 6 to state 7 to state 1 of *Figure 8-14a*. From state 1, if there is a slip, the receiver checks for synchronization in state 2, then state 3, then state 4. If a FAW is found in any one of these states, the receiver immediately is back in synchronization. Note that if the system is initially in alignment but slips, the test for a FAW must fail three times before it is declared to be out of alignment, but with only one find of a FAW, the system is declared back in synchronization.

MORE LINE CODES

Unfortunately, during transmission of standard AMI signals, the pulse density is not adequate for clock recovery without any slips. Special coding techniques have been used to overcome this problem. For example, bits are added to words that contain all 0's. Or, as is recommended by CCITT for Europe for the first order mutiplexer, the even-numbered bits within each 8-bit word are inverted prior to multiplexing. Both methods ensure that enough 1's are in the bit stream for good clock recovery. Let's look at several of these line codes designed specifically to aid clock recovery. *Table 8-2* lists some of the ones used in multiplexing in North America, their bit rates, tolerance and the type of line code. The ones of most interest are the B6ZS and the B3ZS.

***Table 8-2.** AT&T DS Series Line Rates, Tolerances and Line Codes (Format)*

Signal	Repetition Rate (Mbps)	Tolerance (ppm)[a]	Format	Pulse Duty Cycle (%)
DS-0	0.064	b	Bipolar	100
DS-1	1.544	± 130	Bipolar	50
DS-1C	3.152	± 30	Bipolar	50
DS-2	6.312	± 30	B6ZS	50
DS-3	44.736	± 20	B3ZS	50
DS-4	274.176	± 10	Polar	100

[a] Parts per million.
[b] Expressed in terms of slip rate

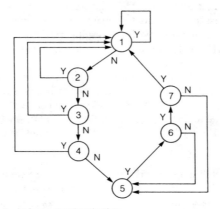

1 : In frame synchronization
2 : Frame code not detected in frame number n
3 : Frame code not detected in frame number n + 1
4 : Frame code not detected in frame number n + 2
5 : Out of frame synchronism (search mode)
6 : Frame code detected in frame number 0 (waiting state)
7 : Frame code detected in frame number 1 (waiting state)

a. State-transition Diagram

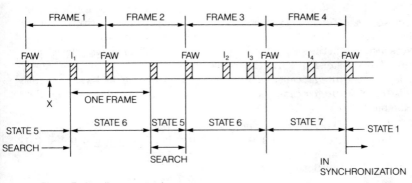

FAW : Frame alignment word
$I_{1,2,3,4}$: Information bits that due to their random nature happen to "imitate" the FAW

b. Synchronization example

Figure 8-14. Frame Alignment
(Courtesy of McGraw-Hill Book Co.)

Binary N-Zero Substitution (BNZS)

The technique used to solve the pulse density problem is to use special coding to assure that sufficient 1's are always available to provide accurate timing. The special code patterns are inserted into the transmitted bit stream. At the receiver, the presence of the special code is detected and the signal adjusted to the original information.

One way of assuring that there are large numbers of ones in any transmitted bit stream is to detect long strings of zeros and substitute for them code patterns containing mostly ones. At the receiver these code patterns are recognized and the inserted pattern is replaced with 0's. Such algorithms are called Binary N-Zero Substitution schemes or BNZS.

B6ZS and B3ZS Codes

One special coding used in the Bell System's T2 transmission lines (the DS-2 signal) is called Binary 6-Zero Substituion (B6ZS). In this special coding, the receiver detects where the substitution has occurred by detecting a bipolar violation. Bipolar violations are illustrated in *Figure 8-15*.

In a normal bipolar code, shown in *Figure 8-15b*, the pulses representing 1's are alternately positive and negative around zero. Zero is represented by 0 and 1 is represented by a + or − sign which indicates the polarity of the pulse. For the normal bipolar code, the + and − signs alternate in the code with 0's placed as they occur in the input stream. Bipolar violations are detected when the alternating pattern of + and − signs is interrupted; that is, if two + or two − signs occur in sequence. Three examples of violations are shown in *Figure 8-15c, d* and *e*.

In the B6ZS code, whenever a string of six zeros occurs, a special code shown in *Table 8-3* is substituted. The code pattern substituted depends on the polarity of the 1 pulse immediately preceding the string of six zeros. Note also that in the substituted code, there is a built-in bipolar violation caused by the substitution of the second and fifth pulses. When the second pulse is substituted it causes a +0+ or a −0− violation, and when the fifth pulse is substituted it causes a −0− or a +0+ violation. Here are specific examples:

1. The 1 pulse preceding six zeros is a +
    Binary Code: 0 1 0 1 <u>0 0 0 0 0 0</u> 1 1 <u>0 0 0 0 0 0</u> 1 0 0 1
    Representation: 0 − 0 <u>+ 0 + − 0 − +</u> − <u>+ 0 + − 0 − +</u> − 0 0 +
    with Substitution:

2. The 1 pulse preceding six zeros is a −
    Binary Code: 0 1 1 1 <u>0 0 0 0 0 0</u> <u>0 0 0 0 0 0</u> 1 1 0 1 0 0 1
    Representation: 0 − + − <u>0 − + 0 + −</u> <u>0 − + 0 + −</u> + − 0 + 0 0 −
    with Substitution:

Table 8-3. *B6ZS Rules*

1 Pulse Polarity	Special Code
+	0 + − 0 − +
−	0 − + 0 + −

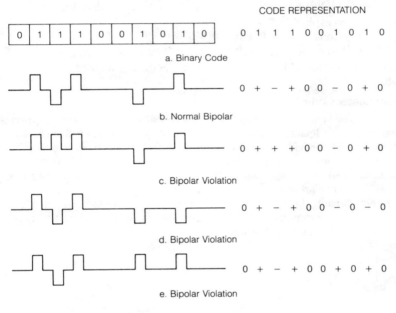

CODE REPRESENTATION

| 0 | 1 | 1 | 1 | 0 | 0 | 1 | 0 | 1 | 0 |

0   1   1   1   0   0   1   0   1   0

a. Binary Code

0   +   −   +   0   0   −   0   +   0

b. Normal Bipolar

0   +   +   +   0   0   −   0   +   0

c. Bipolar Violation

0   +   −   +   0   0   −   0   −   0

d. Bipolar Violation

0   +   −   +   0   0   +   0   +   0

e. Bipolar Violation

Figure 8-15. *Examples of Bipolar Violation*

HDB3

A somehwat different special code substitution is recommended by the CCITT in Europe. It is also a BNZS code called High-Density Bipolar Three-Bit substitution (HDB3). This technique replaces strings of four zeros with the substitute code shown in *Table 8-4*.

Table 8-4. *HDB3 Rules*

1 Pulse Polarity	*Special Code	
	Odd	Even
+	0 0 0 +	− 0 0 −
−	0 0 0 −	+ 0 0 +

*Code Substitution governed by the number of 1's that have occurred since the last substitution.

Because four zeros are replaced (greater than 3), the code is called HDB3. Its substitution rules depend on how many 1's have occurred in the PCM bit stream since the last substitution as well as the polarity of the 1 pulse that is just ahead of the string of zeros. Here are examples:

A. Plus, odd; minus, odd
    Binary Code: 0 1 0 1 1 <u>0 0 0 0</u> 1 0 1 0 1 <u>0 0 0 0</u>
    Substitution: 0 + 0 − + <u>0 0 0 +</u> − 0 + 0 − <u>0 0 0 −</u>

B. Plus, even; plus, odd
    Binary Code: 0 1 0 1 <u>0 0 0 0</u> 0 1 0 1 1 <u>0 0 0 0</u>
    Substitution: 0 − 0 + <u>− 0 0 −</u> 0 + 0 − + <u>0 0 0 +</u>

Note that a bipolar violation occurs in the last bit of the substituted code.

Manchester Coding

The systems discussed above use repeating code patterns to provide adequate pulse densities for good timing recovery and freedom from dc wander, and to provide error indications. A code that has these benefits, but uses only two levels for binary data is the Digital Biphase, or Diphase, or Manchester code. It uses the phase of a square wave signal to indicate a 1 or a 0 as shown in *Figure 8-16b* and *c*. A 0 has an opposite phase waveform from a 1. Every signaling interval contains a zero crossing to provide a good reference for timing recovery and every interval contains an equal amount of positive and negative level for no dc wander.

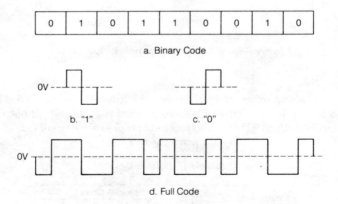

Figure 8-16. Manchester Code

WHAT HAVE WE LEARNED?

1. Digital channel banks produce the PCM signal and multiplex it with many other channels onto a common transmission link.
2. Multiplexers serially arrange PCM signals in time slots with a set number of channels per frame.
3. Multiplexers can have from 24 channels per frame to 4,032 channels per frame and have bit rates from 64 Kbps to 274.176 Mbps.
4. Regenerative repeaters retrieve the PCM signal and reconstruct the pulses.
5. Long strings of binary 1's or 0's can cause dc wander and signal transmisson errors.
6. Receivers must provide clock recovery for data retrieval based upon the received signal.
7. Long strings of 0's in the PCM bit stream can cause a loss of clock and a slip or loss of synchronization.
8. Special line codes are substituted in the transmitted signal to make sure that a proper number of 1's appear in the bit stream to maintain proper timing.

Quiz for Chapter 8

1. TDM carrier systems are growing in number because:
  a. they are easy to install and maintain.
  b. their error rates can be made very low.
  c. they can use low cost large-scale integrated circuits.
  d. they can carry signals other than voice.
  e. all of the above.

2. The noise generated by TDM carrier systems is _____ that generated by analog carrier systems.
  a. greater than
  b. about the same as
  c. a little less than
  d. much less than

3. One section in a D-1 channel bank handles _____ voice channels.
  a. 12
  b. 20
  c. 48
  d. 96
  e. 24

4. Functions performed by D-type channel banks are:
  a. analog to digital conversion.
  b. time division multiplexing.
  c. dialed number translation.
  d. encoding and decoding.
  e. all except c. above.

5. Data sent by a D2 channel bank are at _____ million bits per second.
  a. 1,000
  b. 2.048
  c. 44.736
  d. 1.544

6. Multiplexer systems used in the public network are of the _____ type.
  a. synchronous
  b. plesiochronous
  c. asynchronous
  d. isochronous

7. Pulse stuffing is used in TDMs to:
  a. correct pulses with insufficient height.
  b. add signaling pulses to the data.
  c. provide timing information to the receiver.
  d. equalize the transmission rate between input channels.

8. The Frame Alignment format in the DS-1 signal contains:
  a. 1 bit.
  b. 4 bits.
  c. 7 bits.
  d. 8 bits.

9. Synchronization of digital switches in the North American telephone network is done using a _____ technique.
  a. independent
  b. master-slave
  c. mutual conductance
  d. plesiochronous

10. The line coding scheme used for the DS-1 signal is:
  a. non-return-to-zero.
  b. binary 3-zero substitution.
  c. ternary.
  d. bipolar.

Modems — Telephone Service for Computers

ABOUT THIS CHAPTER

This chapter is about modems— the equipment that provides the interface between the outputs of all types of digital equipment and telephone lines. Material covered includes the types of signals transmitted, the modulation techniques, the timing requirements, the signal conditions, and the function and operation of the electronic devices that provide the interface.

WHAT IS A MODEM, AND WHY IS IT REQUIRED?

Computers, like people, must communicate with each other. They pass information back and forth as electrical signals called non-voice or data. The transmission link for the information varies depending on the physical location. If the machines are located in the same room or the same building, they probably are connected together directly. For longer distances, they are connected together through the telephone network. Increasing amounts of non-voice information flow over the public network between machines each day. Let's look at this connection more closely.

Figure 9-1 shows two pieces of digital equipment (data terminal equipment) interconnected to communicate digital information between them. The piece of equipment in *Figure 9-1a* is sending the digital data to the equipment in *Figure 9-1b* located a long distance away. Communication can occur in the same fashion in the reverse direction. A telephone line is the transmission link between them.

As shown in *Figure 9-1a*, the digital equipment outputs bits in parallel to represent the data. The parallel output is converted to a serial bit stream by the transmitter portion of a Universal Asynchronous Receiver Transmitter (UART) so that the information can be sent in serial form over a single line. The least significant bit (LSB) is sent first and the most significant bit (MSB) is sent last. The serial bit stream can't be transmitted directly, but must be changed to tones that can be sent over the telephone line. A modem provides the conversion and coupling to the telephone line. The word modem is a combination of the words MOdulator and DEModulator, the two principal functions accomplished by a modem.

At the receiving end of the telephone line shown in *Figure 9-1b*, the signal is converted back to a serial bit stream by the modem, and the receiver portion of the UART converts the serial bit stream to parallel data for input to the digital equipment. Some digital equipment can accept the serial data directly from the modem; if so, the UART is not needed.

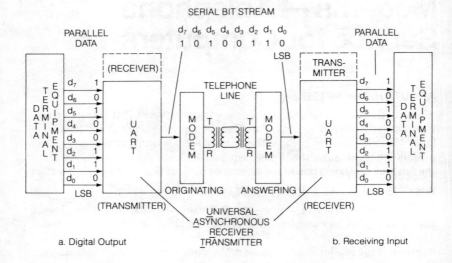

Figure 9-1. *Communicating Digital Data*

A serial bit stream of 8 bits is shown in *Figure 9-1*. It is input to the modem at the sending end, called the originating modem, and is output from the modem at the receiving end, called the answering modem. The format for each character in the bit stream is shown in *Figure 9-2*.

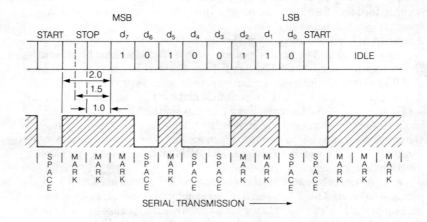

Figure 9-2. *Asynchronous (Start-Stop) Character Format*

Asynchronous Character Format

The bit stream has mark and space intervals in the format. A mark is represented by the presence of some predefined voltage or current level, while a space is represented by a different voltage or current level. Since an indication is needed that a transmission link is established and ready even when data are not being transmitted, the idle state is represented by the mark state; therefore, current or tone is present continuously in an idle condition if the circuit is complete. In *Figure 9-2*, a mark represents the binary 1 and a space represents the binary 0 in the digital data as the bit stream is formatted and transmitted.

An asynchronous character format has additional characteristics that provide a timing reference for the receiver. Before the code representing a character of the data (which may be a letter, number, command or special symbol), there must be a start bit. It is one space interval. After the character code, there must be a stop bit. It varies in length from one interval to two intervals depending on the code and the equipment. The start interval indicates the start of the character code and a stop interval provides the receiver circuits a reference for detection of the beginning of the next character. The format allows the character code itself to be of any practical length, but typically is from 5 to 8 bits long.

To summarize, asynchronous operation requires that the transmitter (sending machine) and the receiver maintain within themselves free-running clocks with the same nominal frequency which is accurate to within certain tolerances (which depend on the maximum data rate to be transmitted). Each encoded symbol, usually called a character or a byte, is preceded by a known state of the input signal, called the rest or idle state, followed by a reference or synchronizing signal called a start bit. Besides the data rate, additional parameters upon which the sender and receiver must agree are the length of the signal interval (sometimes called the bit time), the number of signaling intervals per symbol (bits per byte or per character), and the minimum length of the stop bit or idle interval which must precede each new character. Typical stop bit intervals are 1.0, 1.42, 1.5 and 2.0 bit times as previously indicated. A 5-bit code like that shown in *Figure 9-3a*, is commonly called a Baudot code, after Emile Baudot who devised the first constant length code for teleprinters in 1874. It used a 1.42 stop interval for a total of 7.42 intervals per character.

5-Bit and 7-Bit Codes

To illustrate what the serial bit stream would look like at the input to the originating modem, the letter S is shown in *Figure 9-3* represented in two different codes—the 5-bit CCITT alphabet No. 2 code and the 7-bit American Standard Code for Information Exchange (ASCII). Note that the stop interval is shown as 1.5 for the CCITT code and 2.0 for the ASCII. In many cases, an eighth bit, called the parity bit, is appended to the ASCII to use for detecting errors that may occur in transmission.

Parity may be odd or even in a particular system. If even parity is used, the parity bit is set to a 1 if needed to make the total number of ones in the 8-bit code an even number. Conversely, if odd parity is used, the parity bit is set to a 1 if needed to make the total number of ones in the 8-bit code an odd number.

Asynchronous Operation

The communication between the data terminal equipment shown in *Figure 9-1* using the format of *Figure 9-2* is said to be asynchronous. That is, there is no signal within the character format that conveys the timing between transmitter and receiver for each bit. The start bit is used as a synchronizing signal by transmitter and receiver for *each* character, but as the receiver inputs the code, an oscillator in the receiver determines the input timing independently of the data. The timing oscillator frequencies of the transmitter and receiver must stay within certain limits so that the signals at the receiver do not drift beyond the correct sampling points at the receiver.

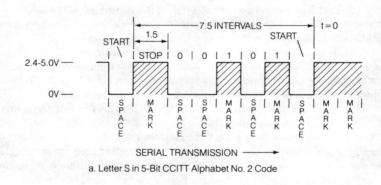

a. Letter S in 5-Bit CCITT Alphabet No. 2 Code

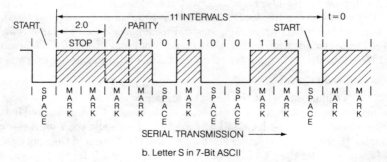

b. Letter S in 7-Bit ASCII

Figure 9-3. *5- and 7-Bit Characters in Asynchronous Format*

Figure 9-4 illustrates the asynchronous operation and shows that even though the transmitter timing clock drifts with respect to the receiver clock, the sampling points of the receiver still can detect the correct bit pattern from the transmitted pulses. The first sample of the data, identified as A in *Figure 9-4*, detects the beginning of the start interval. A timing circuit in the receiver, set for 50% of the average clock time of the transmitter, samples the incoming data again at sampling point B. If sampling point B continues to detect a space level, then the start bit is considered valid and the receiver continues to sample to accept the character bits. Every interval is sampled at the 50% interval point. The timing is determined by the receiver clock. Sampling points C through J recover the 7-bit character code and the parity bit, and samples K and L recover the stop interval. The interval between characters may be of any length. When the next start bit is detected, the timing clocks are reset to synchronize the timing and the sampling process repeats.

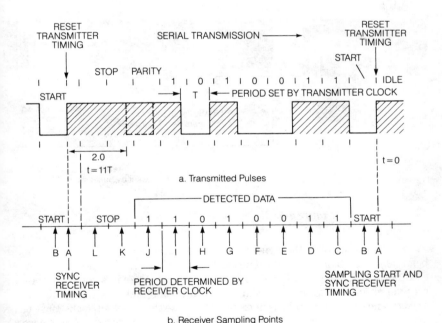

Figure 9-4. Asynchronous Operation of Transmitter and Receiver

Synchronous Operation

Synchronous operation is truly synchronous because a clock signal is transmitted with the data to maintain the transmitter and receiver in continuous synchronization. The format for the digital data is as shown in *Figure 9-5*. Each of the blocks in *Figure 9-5* represents an 8-bit character because a 7-bit ASCII character with parity was selected for the example. There are no start or stop bits associated with each synchronous character. All the bits for a group of characters are sent one after the other in what are called "blocks of data". As a result, the timing of transmitter and receiver must be more accurately synchronized than for the asynchronous transmission.

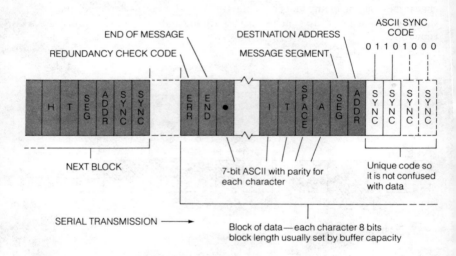

Figure 9-5. Synchronous Message Format

To accomplish this synchronization, special codes are included at the beginning of each block of data. The electronic circuits in the receiver constantly check the incoming data for this pattern. When it is detected, the receiver assumes that the next character is data, and that data will continue until an end of message code is detected. (Because the receiver clocks must be derived from the input data stream, synchronous modems are generally more expensive than asynchronous modems.)

There are generally multiple synchronization codes at the beginning of each message in case the first code is lost because of transmission problems. The length of the block usually is dependent on a buffer memory that stores the block of data before transmission. This buffer is not necessary in asynchronous transmission because each character is transmitted as soon as it is produced by the digital equipment. But for synchronous transmission, several characters are stored, then all are transmitted at a regular and constant rate of so many bits per second. For the same reason, buffers usually are required at the receiving end.

The synchronous data format shown in *Figure 9-5* is not the only one possible, but it demonstrates the principle. Some synchronization techniques have sync codes inserted at particular time intervals. Some have special framing formats. With the advent of high-speed integrated ciruits and very large-scale integration (VLSI), some of the synchronization techniques are changing. More recent circuit designs detect the transitions of the received digital data and continually adjust the receiver clock for even more accurate synchronization.

Isochronous Operation

Isochronous operation is a mix between asynchronous and synchronous operation. *Figure 9-6* shows the character format. Individual characters are framed with a start and stop bit as in asynchronous operation, but the intervals between characters are time controlled. The time interval may be of any length so long as it is restricted to multiples of one character time.

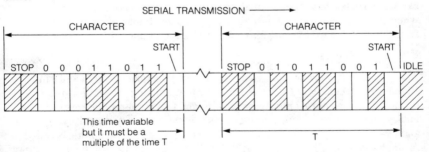

Figure 9-6. Isochronous Character Format

MODULATION AND DEMODULATION

Why is it that the digital signal output from a computer cannot be connected directly to a telephone line? The reason basically has to do with the amount of bandwidth available to carry signals over a single telephone channel, which in turn is related to the cost of the channel. The lowest-cost channels currently available are voiceband channels, which have been designed to carry voice signals produced by telephone sets, and whose bandwidth extends from about 300 Hz to about 3,400 Hz. The means the channel cannot pass signals of very low frequency (like direct current), nor signals of very high frequency (above 3,400 Hz). The signals used in computers are usually unipolar signals and the change (transition) from the 0 level to the 1 level is very fast (at a high frequency). As a result, the computer signals contain significant frequencies below 300 Hz (even a dc component) and frequencies well above the 3,400 Hz limit of a VF telephone channel. The fact that the channel will not carry these frequencies leads to the necessity of transforming the digital waveform to a signal that is compatible with the channel and its bandwidth. The means used for the transformation is called modulation.

Modulation

Modulation is the process of changing some property of an electrical wave (called the carrier) in response to some property of another signal (called the modulating signal). In the case of transmission of data signals over the telephone channel, modulation involves changing some property of an alternating current wave carrier of between 300 and 3,400 Hz in response to a binary (1 or 0) signal from a computer. The properties of the carrier that are available to be changed are the amplitude, frequency, and phase as illustrated in *Figure 9-7*. Each of these methods is used in modems. Let's look at them in more detail.

Amplitude Modulation

Amplitude modulation is the process of changing the amplitude of the carrier in response to the modulating signal. As illustrated in *Figure 9-7b*, when the modulating signal is a binary signal, the amplitude may be varied from 0 for a binary 0 to some maximum value for a binary 1.

Frequency Modulation

Another property of a carrier that can be varied by a modulating signal is the frequency. *Figure 9-7c* shows how shifting the carrier frequency to a lower value represents a binary zero, and to a higher value represents a binary one. This technique is sometimes called Frequency Shift Keying (FSK), and when the frequencies used are in the voice band, the technique is called Audio Frequency Shift Keying (AFSK). The most widely used modems today carry digital signals in the range of 45 to 1,800 bits per second using the AFSK technique.

Phase Modulation

A third property of a carrier that can be varied in response to a modulating signal is the phase. Phase can be visualized as the relative relationship of two waveforms at any given time in their cycle. *Figure 9-7d* illustrates that the phase is shifted for every occurrence of a one bit, but the phase is not shifted for a zero bit. This technique is called Phase Shift Keying (PSK). The phase of the signal sent over the transmission medium usually is not measured absolutely, but rather is measured relative to the phase of the wave during the previous bit interval.

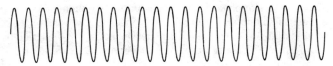

a. Carrier

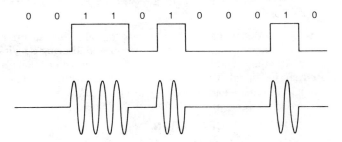

b. Amplitude Modulation

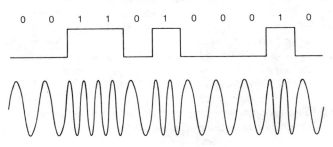

c. Frequency Modulation

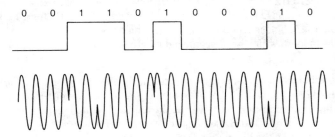

d. Phase Modulation

Figure 9-7. *Types of Modulation*
(From D. Doll, Data Communications; Facilities, Networks and System Design, *John Wiley & Sons, 1978, Copyright© 1978 by John Wiley & Sons, Inc., Reprinted by permission of John Wiley & Sons, Inc.)*

Transmission of phase information on the telephone network presents some challenges. The human ear is relatively insensitive to the phase of speech or music sounds; therefore, the telephone network was not designed to preserve carefully the phase relationships of signals sent through it. High-speed modems which use phase modulation to carry data usually have circuits in them which compensate for disturbances or nonlinearities in the telephone network to help restore the phase linearity of the received signal.

Demodulation

As the carrier is modulated to carry the digital signal, so it must be demodulated to recover the digital signal. Detector and filtering circuits sensitive to either amplitude, frequency or phase recover the 1's and 0's from the modulated carrier signal. Threshold bias and level shifting circuits restore the digital signal to common logic levels or special interface transmission levels specified by the digital signal protocols for interfacing.

ASYNCHRONOUS MODEM OPERATION

Figure 9-8 shows the frequency modulation schemes used in most low-speed (up to 300 bps) asynchronous modems. In *Figure 9-8a*, a center frequency carrier (1,170 Hz) is frequency shifted to 1,270 Hz for a 1 and frequency shifted to 1,070 Hz for a 0. Every time the serial bit stream for a character is input to the originating modem, the output is a continuous alternating signal of 1,070 Hz or 1,270 Hz depending on whether the input is a 0 or a 1. This scheme is used for simplex transmission (transmission in one direction only) and for half-duplex transmission (transmission in both directions, but only in one at a time).

If full-duplex transmission (transmission in both directions at the same time) is required, then the modulation is accomplished by dividing the available bandwidth into two bands as shown in *Figure 9-8b*. The lower band carries data in one directon and the upper band carries data in the other direction. The lower band center frequency is 1,170 Hz and the freqency is shifted to 1,270 Hz for a 1 and to 1,070 Hz for a 0. The upper band center frequency is 2,125 Hz and the frequency is shifted to 2,225 Hz for a 1 and to 2,025 Hz for a 0. These frequency pairs are the ones used in the Bell System 103 series modem which has become a de facto (by use) U.S. standard.

System Interconnection

When both modems are operated in the full-duplex mode, *Figure 9-9a* shows the block diagram of the interconnection. The transmitter at the originating modem and the receiver at the answering modem operate at the center frequency of 1,170 Hz, while the transmitter at the answering modem and the receiver at the originating modem operate at the center frequency of 2,125 Hz. *Figure 9-9b* shows the relative signal levels versus frequency for the two signal channels. *Figure 9-9c* shows the frequencies for a similar full-duplex CCITT system used in Europe.

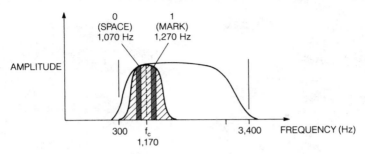

a. Simplex or Half-Duplex Transmission Modulation

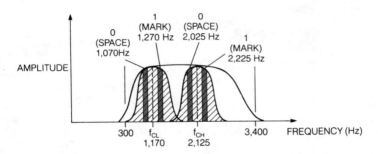

b. Full-Duplex Transmission Modulation

Figure 9-8. *AFSK Modulation*

1200 bps Asynchronous Modem

Another method of dividing the available frequencies is illustrated in *Figure 9-10*. In this case, the modem can carry asynchronous data at 1,200 bits per second, but in only one direction at a time (half-duplex operation). The higher data rate requires a wider bandwidth so the two frequencies for 0 and 1 must be spaced further apart; therefore, there is not enough bandwidth in the channel for full-duplex operation. The modems using this scheme are called 202-compatible, since the Bell System 202 modems established this scheme. Enough bandwidth is available in the channel with the 202 frequencies to allow a single-frequency signal at 387 Hz to be transmitted in the reverse direction while the 1,200 bps data are being sent forward. This reverse channel is used mainly to transmit a continuity tone for the receiver to tell the transmitter that the circuit is established, but some data may be sent using on-off keying of the tone. Modems of modern design using low-cost electronics to perform more advanced modulation techniques now carry 1,200 bps data in full duplex mode, and some offer compatibility with either 300 bps 103-type signals or 1,200 bps data.

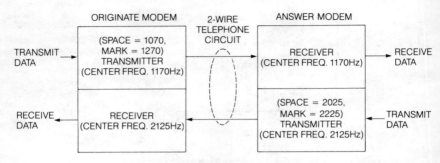

a. U.S. Frequency Assignments

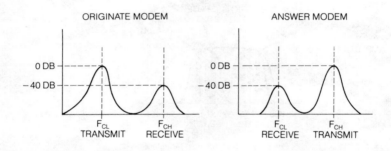

b. Transmit and Receive Levels

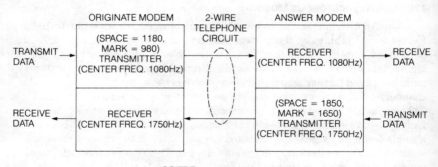

c. CCITT Frequency Assignments

Figure 9-9. *Low-Speed Asynchronous Full-Duplex Modem Frequency Assignments and Levels*
(*Courtesy of Digital Press. Reprinted from J.E. McNamara,* Technical Aspects of Data Communication, *Second Edition, Copyright© 1982 by Digital Equipment Corporation*)

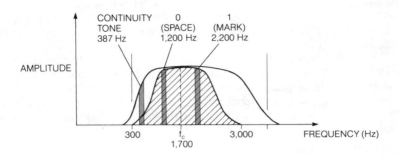

Figure 9-10. Bell 202 Compatible Modem

Bell 103-Type Modem

Figure 9-11 is a block diagram of a 103-type modem. The 2-wire telephone line is terminated in a line matching transformer. The transformer secondary is connected to both the receive section input and the transmit section output, but the received signal has no effect on the transmit section. The transmit output signal could affect the receive section, but the receive bandpass filter prevents the transmit signal from entering the receive section because the transmit and receive frequencies are in different bands. The receive bandpass filter also rejects noise and spurious frequencies riding on the receive signal from the telephone line. The limiter eliminates amplitude variations.

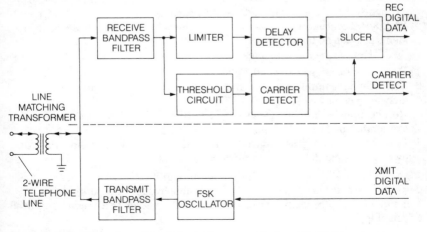

Figure 9-11. Low-Speed Asynchronous Modem Block Diagram

A delay detector provides a delayed sample of the signal, compares it to the receive signal, and gives an output which is proportional to the difference in frequency. The slicer circuit clips the top and bottom of the detected signal and produces a digital signal at the output with the proper digital voltage levels for 1 and 0.

When data are to be transmitted, the digital serial bit stream is applied to a frequency shift oscillator which produces the audio frequency shifted tones representing the ones and zeros. This signal is filtered by the bandpass filter to remove spurious harmonics (particularly those within the receive passband) and is passed through the line matching transformer to the telephone line.

Acoustical Coupler

In *Figure 9-11*, the modem is shown to be connected directly by wires to the telephone line. This is the preferred method; however, some modems are acoustically coupled to the telephone handset. The coupler has a microphone and speaker which interfaces with the handset receiver and transmitter, respectively, to couple the transmitted and received signals by sound waves. Thus, there are no wire connections between the modem and telephone line. A telephone call is made to the destination and a connection established the same way as for a conversation before the handset is placed in the coupler.

Advances in Modern Modems

Original modems using circuitry like *Figure 9-11* occupied a volume of 600 cubic inches (9,832 cubic centimeters) in a package 10 × 10 × 6 inches (25.4 × 25.4 × 15.2 centimeters). Using the latest integrated circuits, modems are now manufactured occupying only one-tenth of that volume, 65.3 cubic inches (1,070 cubic centimeters), in a package 5.5 × 9.5 × 1.25 inches (14 × 24.1 × 3.2 centimeters). Besides the smaller physical size, these modern modems provide added performance features.

Here are some examples:

1. Provide data rates at 300 and 1,200 bps.
2. Dial telephone numbers automatically for origination and automatically answer incoming calls.
3. Return alphabetic characters to the data terminal equipment to report on the condition of the telephone line and the call in progress.
4. Detect the answer tone from a distant modem automatically and adjust the data rate to match the distant modem's data rate.

The 10-fold reduction in size and the significant increase in performance and "intelligence" is made possible by using microprocessor techniques and by reducing the electronics of the modem to a single integrated circuit chip.

One-Chip I/C Modem

A block diagram of such a one-chip modem interfaced to a telephone line is shown in *Figure 9-12*. The particular device shown is a Bell 103 compatible FSK modem, the TMS99532 made by Texas Instruments. (Texas Instruments also makes a one-chip modem, the TCM3101, that operates similarly and meets Bell 202/CCITT V.23 specifications.) The telephone line is connected to the TXA (Transmit Analog) and RCVA (Receive Analog) lines. The binary input from attached digital equipment that delivers formatted serial bit streams is connected to the XMTD (Transmit Digital Data). The received data converted to binary format are output on the RCVD (Receive Digital Data) line. The frequency arrangement is as shown in *Figure 9-9a*. In the answer mode, digital data are received in the low-band frequencies and transmitted in the high-band frequencies. The reverse is true in the originate mode.

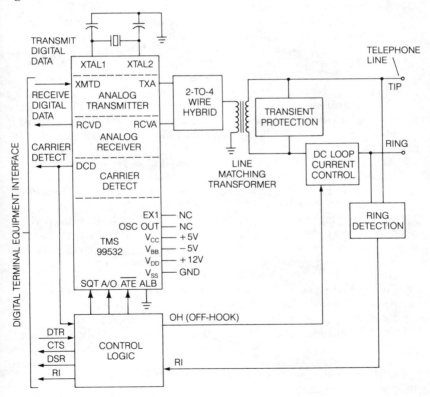

Figure 9-12. One-Chip Modem

The modem chip must be interfaced to the telephone line through a 2-to-4-wire hybrid and a line matching transformer. Transient protection, on-hook/off-hook control, and ring detection must be provided externally as shown. (There are commercial circuits available to do these functions. An example is the Cermetek CH1810). An external crystal is used to provide accurate control of the frequency of the modem's on-chip oscillator for the timing. Power at +12V, +5V, −5V and ground must be supplied to the modem.

Establishing a Connection

To understand the modem operation in more detail and the interconnection signaling that takes place, let's go through the steps that occur when two modems are put in operation on the ends of a telephone line. (These steps are often called "handshaking".) Here are the steps:

1. Each modem is connected to a terminal and power supplied. The originating modem has the A/O control line set at a 1 and the answering modem has it set at 0. No connection is made to EX1 and OSC OUT. The ALB (Analog Loop Back) pin must be grounded for normal operation.
2. The $\overline{ATE}$ control line for both modems is set at 1 (inactive) since Bell 103 type modems use the 2,225 Hz mark frequency as the answer tone. If CCITT interconnections were in force, the $\overline{ATE}$ line would be under logic control to turn on a 2,100 Hz answer tone according to CCITT requirements.
3. The SQT input is under logic control. When a 1 (active), the transmitter is off and no signal is output on TXA. Under idle conditions, it is active and the transmitter is off.
4. At the originating modem, a telephone call is placed to the answering modem number and a ring signal appears at the answering modem.
5. The answering modem detects the ring signal and is enabled by assertion of DTR (Data Terminal Ready) to go off-hook.
6. The answering modem waits 2 seconds (specified by the FCC to provide a billing delay) and, through SQT, turns on its transmitter to send the answer mode carrier (mark frequency 2,225 Hz) to the originating modem.
7. The originating modem recognizes the answer mode signal and places a 0 (active) on $\overline{DCD}$, the carrier detect line.
8. The originating modem turns on its transmitter by placing a 1 on SQT as a result of the $\overline{DCD}$ carrier detect. It sends an originating mark signal (1,270 Hz) down the line. It also starts a 200-350 millisecond CTS (Clear To Send) time delay to prevent the digital equipment from sending data.
9. The answering modem receives and recognizes the originating modem mark signal and places a 0 on its $\overline{DCD}$ carrier detect line. Now both modems are sending carriers in both directions.
10. The answering modem, through the $\overline{DCD}$ carrier detect, puts a 1 (active) on the CTS line to the digital equipment and it's digital equipment is clear to send data communications.

11. At the originating modem the CTS time delay expires, the CTS signal is changed to a 1, and the digital equipment connected to the originating modem is clear to send data communications.

Several techniques are available for disconnect control. A common one negates the DTR signal when sending is complete, which causes the modem to go off-hook, terminating the call.

The connection was discussed for asynchronous modems. Similar connections are required for synchronous modems except more signals are required for timing and transmission rate control.

STANDARDS FOR DIGITAL EQUIPMENT INTERFACE

The modem must exchange control signals with whatever terminal or business machine is connected to it. The frightening possibility of large numbers of mismatched control signals and connectors between modems and data terminals has resulted in the establishment of several standards to define the physical and electrical parameters between modems, called data communications equipment (DCE), and terminals, called data terminal equipment (DTE). The most common U.S. standard, administered by the Electronic Industries Association, is called EIA RS-232C. The European standard, administered by the CCITT, is known as CCITT Recommendation V.24.

The basic signals required for sending and receiving data, controlling the modem from a terminal, and passing status information back to the terminal from the modem, along with their standard designations and pin assignments for the standard connector are shown in *Table 9-1*.

Synchronous modems require more interface leads than asynchronous modems because the clocking signal for both the send and receive data must be accommodated, and the standards specify interface leads for selecting alternative data rates (in case of excessive error rates) and a signal from the modem indicating a high probability of an error in the received data.

In addition to specifying the standard pin designations and connection, the EIA RS-232C/CCITT V.24 standards specify signal conditions and levels that must occur on the interchange. For example, the voltages specified by the EIA standard are $+3V$ to $+25V$ for a space (zero) and $-3V$ to $-25V$ for a mark (one). The circuit implementation must meet certain electrical specifications other than voltage, such as withstanding dead shorts between conductors without damage, and cannot exceed maximum values for load resistance and effective capacitance. To aid in meeting the line interface requirements, special line drivers and line receivers have been designed in integrated circuit form. If a line is to receive data that are being sent under RS232 standards, a SN75189 line receiver can be placed in the line. If a line is to transmit signals at RS-232 standards, then a SN75188 line driver can be used to drive the line. The interconnection is shown in *Figure 9-13a*.

Table 9-1. *Internationally Accepted Assignments for DTE/DCE Interface Using a 25-Pin Connector*

Pin	CCITT	EIA	Circuit Name
1	101	AA	Protective Ground (not always used)
2	103	BA	Transmitted Data
3	104	BB	Received Data
4	105	CA	Request to Send
5	106	CB	Clear to Send
6	107	CC	Data Set Ready
7	102	AB	Signal Ground
8	109	CF	Received Line Signal (Carrier Detector)
9	Often used for modem power test point—do not connect		
10	Often used for modem power test point—do not connect		
11			
12			
13			
14			
15	114	DB	Transmit Signal Element Timing—DCE Source (Synchronous Modems Only)
16			
17	115	DD	Receive Signal Element Timing—DCE Source (Synchronous Modems Only)
18			
19			
20	108	CD	Data Terminal Ready
21			
22	125	CE	Ring/Calling Indicator
23			
24			
25			

(Courtesy of Digital Press. Reprinted from J.E. McNamara, Technical Aspects of Data Communication, *Second Edition, Copyright© 1982 by Digital Equipment Corporation)*

The EIA RS-232C/CCITT V.24 standards specify an unbalanced or single-ended interface for each control circuit *(Figure 9-13a)*. Such circuits have problems with noise immunity and ability to drive long lengths of cable. New standards, designated EIA RS-422 and EIA RS-423 specify both a balanced and unbalanced interface as shown in *Figure 9-13b* and *c*. In addition to different termination configurations, these standards specify limits for the rise time of the digital signals to lower the probability of crosstalk on adjacent circuits and improve the maximum signaling distance from 15 meters to over 1,000 meters for data rates under 100,000 bits per second. New EIA standards (RS-485) that are very similar to RS-422 cover the case where up to 32 drivers and 32 receivers may be on the same line.

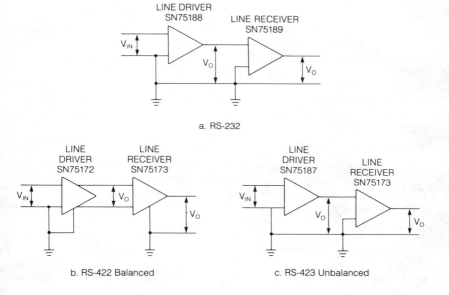

Figure 9-13. *Line Driver-Line Receiver Interfaces*

ERROR DETECTION AND CORRECTION

Although the subject of error detection and correction has
nothing to do directly with the subject of modems, whenever digital data are
transmitted over telephone channels using modems, errors are sure to occur.
The computer systems and terminals that produce and consume the data must
take measures to detect and, if possible, correct the data errors.

Recent modem developments combined with the use of
microprocessors have allowed the data to be buffered, error control applied,
and checked by the receiver. If the receiver detects an error, it automatically
requests retransmission of the erroneous data.

Error control is applied by adding redundancy; that is, information in
addition to the minimum required to send the original data. The redundant
information is related to the original input in some systematic way so that it
can be regenerated when it is received. Upon receipt, if the regenerated error
control matches that which was sent along with the data, it is assumed the
transmission is error-free.

The required redundancy is provided in different ways. The redundancy to detect errors in long blocks of data is provided by a class of codes call cyclic redundancy check (CRC) codes. The process of generating a CRC for a message involves dividing the message by a polynomial, producing a quotient and a remainder. The remainder, which usually is two characters (16 bits) in length (see ERR of *Figure 9-5*), is added to the message and transmitted. The added information is sometimes referred to as a block check character (BCC). The receiver performs the same operation on the received message and compares its calculated remainder with the received remainder. If they are equal, the probability is quite high that the message was received correctly.

PROTOCOLS

There are also rules for the interaction of communications equipment that usually are implemented through the programming of the data terminal equipment involved rather than being built into the hardware. These are called protocols. Protocols set the rules for grouping bits and characters (framing), error detection and correction (error control), the numbering of messages (sequencing), separating control and data characters (transparency), sorting out receiving and sending equipment (line control), the actions required on start-up (start-up control), and the actions required on shut-down (time-out control). Protocols may be character oriented like IBM's Binary Synchronous Data Transmission (BiSync); they may be byte oriented like Digital Equipment Corporation's Digital Data Communication Message Protocol (DDCMP), or they may be bit oriented like IBM's Synchronous Data Link Control (SDLC).

An example of each of the three protocols is shown in *Figure 9-14*. The character oriented BiSync protocol uses different characters in the character set to indicate the beginning of the heading, the beginning of the message text, and the end of the text. The character indicating the beginning of the message header, called the SOH, causes the accumulation of the redundancy check character (BCC) to be reset. The byte oriented DDCMP protocol begins with a header that gives the type of message, the number of text characters in the message, a message number, destination address, and a Cyclic Redundancy Check (CRC). The message text follows and the frame is ended by another CRC. The frame of the bit oriented SDLC protocol begins and ends with the unique bit sequence 01111110. The bits shown in the Frame Check Sequence (FCS) are the CRC of the frame. The most important property is that the message is always transparent; that is, it may contain any sequence of bits, without being mistaken for a control character.

WHAT HAVE WE LEARNED?

1. Modems change digital signals into signals that can be handled over standard telephone lines.
2. Digital data are formatted in particular serial sequences in order to transmit them asynchronously or synchronously.

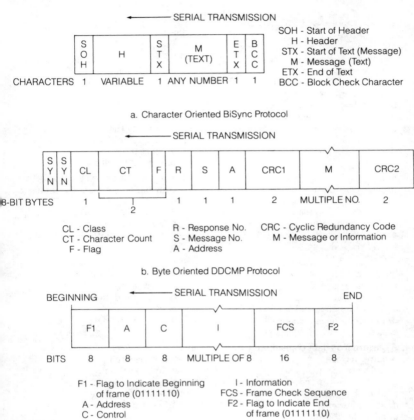

a. Character Oriented BiSync Protocol

b. Byte Oriented DDCMP Protocol

c. Bit Oriented SDLC Protocol

***Figure 9-14.** Protocol Examples*

3. When transmitting digital data over telephone lines, there is an originating modem on one end and an answering modem on the other.
4. Modems may have the capability for full-duplex, half-duplex, or simplex communication.
5. A modem modulates a carrier to transmit digital data.
6. A modem demodulates a carrier to convert transmitted data to the original digital data.
7. EIA standards for the U.S. and CCITT standards for Europe set the signal levels and pin connections for interfacing modems to data terminal equipment.
8. Protocols are rules for digital communications that govern how data terminal equipment is programmed to accomplish the communications.

Quiz for Chapter 9

1. Modems are required to connect computers to telephone lines because:
 a. the telephone network bandwidth is too high.
 b. the telephone network will not pass direct current.
 c. telephone company rules require them.
 d. none of the above.

2. What kind of modulation is used in modems?
 a. phase modulation.
 b. frequency modulation.
 c. amplitude modulation.
 d. pulse modulation.
 e. all of the above.
 f. all except d.

3. The most common technique for binary data transmission is:
 a. synchronous transmission.
 b. bisynchronous transmission.
 c. asynchronous transmission.
 d. plesiochronous transmission.
 e. none of the above.

4. Asynchronous data transmission requires a clock:
 a. at the transmitter end.
 b. at the receiver end.
 c. at both ends.
 d. at neither end.
 e. all of the above.

5. The data transmission code most widely used in the U.S. is:
 a. Fielddata.
 b. Baudot.
 c. EBCDIC.
 d. ASCII.
 e. CCITT #5.

6. 103-type and 202-type modems use what modulation scheme?
 a. AFSK
 b. PSK
 c. MFM
 d. SSB-AM
 e. TDMA

7. The difference between asynchronous and synchronous transmission is:
 a. the way the synchronization is provided.
 b. the way the beginning of a block of data is detected.
 c. the way the beginning of a character is determined.
 d. all of the above.
 e. none of the above.

8. The parameter that most affects transmission of high-speed modem data is:
 a. phase distortion.
 b. amplitude distortion.
 c. frequency shift.
 d. impulse noise.

9. Error control of data transmission is done by:
 a. retransmission.
 b. adding redundancy.
 c. parity.
 d. cyclic redundancy checks.
 e. all of the above.

10. Protocols may be:
 a. bit oriented.
 b. byte oriented.
 c. character oriented.
 d. any of the above.
 e. none of the above.

Wireless Telephones

ABOUT THIS CHAPTER

This chapter focuses attention on telephones that perform most or all of the functions of the conventional telephone, but are connected with a radio link rather than wired directly. Although the term "wireless" may seem a bit old-fashioned, it is about the only general name for such telephones, since terms like "cordless telephone", "mobile telephone", and even "radio telephone" have come to mean specific types of telephones without wires.

CORDLESS TELEPHONES

The first type of wireless telephone to be discussed is the "cordless" which is used as an extension telephone in homes and businesses. *Figure 10-1* shows that the cordless telephone consists of two parts: a base unit and a portable unit. The connecting wires of a conventional telephone between the portable unit and the base unit are replaced with low-power radio transmissions. The radio link is completed by the transmission of a carrier that is frequency modulated (FM) with the information to be transmitted. The carrier and modulation principles are the same as described for the modem in Chapter 9; the only difference is that the frequency of the carrier is much higher for the cordless telephone.

The cordless telephone is one of two types. For one type, the portable unit has no keypad, and only the voice communications are radio linked. The keypad is on the base unit. The other type has the keypad on the portable unit and all the normal functions of a telephone are radio linked to the base unit.

The telephones are electronic telephones as have been described in this book with the radio frequency transmission and modulating and demodulating electronic circuits added for the radio link. They have pulse generators and/or DTMF generators for dialing, electronic single- or dual-frequency ringers, and electronic speech circuits. Some have special features that allow the telephone to be used as an intercom. Some have speech and/or ringer volume controls and some have special security features to prevent unauthorized use. Most have the redial feature and some have a memory to store several numbers. The cord type electronic telephone usually obtains the operating power directly from the telephone line as has been described earlier; however, the cordless telephone, because of its greater power requirements, usually obtains operating power from the household ac outlet. This is a minor disadvantage since the cordless telephone won't operate if a utility power failure occurs.

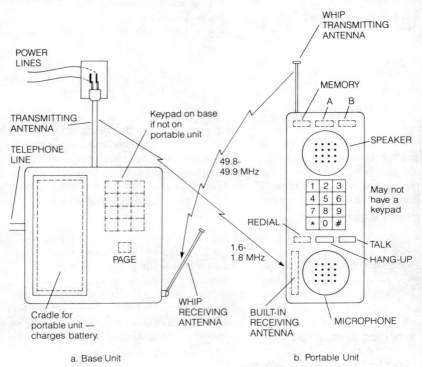

Figure 10-1. Cordless Telephone

Base Unit

As shown in *Figure 10-1*, the base unit connects directly to the telephone line to complete the local loop to the central office. A 2-wire to 4-wire hybrid arrangement couples the local loop to the separate transmit and receive sections in the base unit. The base unit transmits on a carrier frequency in the range from 1.6 to 1.8 MHz and the household electrical wiring is used as the transmitting antenna for the base unit. The nominal 1.7 MHz frequency modulated signal is fed from the base unit transmitter to the ac line through capacitors which block the line current from the base unit transmitter while passing the 1.7 MHz output to the line. This use of the house wiring as an antenna is not unique to cordless telephones; it also is used for wireless intercoms. This method provides good reception within and near the house, as well as outside near power lines that are on the same side of the utility company's distribution transformer as the house circuit. This may include a neighbor's house wiring; thus, the potential exists for interference if that person also has a cordless telephone; more about this later.

Portable Unit

An internal loopstick antenna (like that used in standard radio receivers) in the portable unit receives the nominal 1.7 MHz transmission from the base unit over a range from 50 to 1,000 feet. The range depends not only on the manufacturer's design, but also on such things as whether the house wiring is enclosed in metal conduit and whether foil-backed insulation is used in the walls. The ringing or voice signal is recovered by demodulation and drives the speaker in the portable unit. The portable unit is powered by a battery which is recharged when placed in a receptacle in the base unit.

The portable unit is usually in a standby mode which corresponds to the on-hook condition of a telephone set. When the ringer sounds, the user operates a talk switch which turns on the transmitter in the portable unit. This transmitter transmits on a frequency in the range of 49.8 to 49.9 MHz and outputs the signal on the whip antenna. (Since the whip is used only for transmit, it can be collapsed out of the way when the portable unit is on standby. If the portable unit is only for voice transmission, it may have an internal antenna and its range is shorter.) A similar whip antenna on the base unit receives the FM signal from the portable unit, demodulates it and applies the off-hook signal to the telephone local loop.

When the user dials the number for outgoing calls, the dial pulses produce tones which modulate the carrier for transmission to the base unit. The base unit recovers the tones by demodulation. If DTMF service is used, the tones are sent on the telephone line. If pulse service is used, the tones are converted to pulses and the telephone line is pulsed. When the connections between calling and called parties are established, both transmitters and receivers operate at the same time to permit two-way conversation.

Frequencies

Although a few cordless telephones use the same frequency for transmission in both directions, most use two different frequencies in the ranges given above. Unless multiplexing is used, the use of a single frequency provides only half-duplex (one-way-at-a-time) transmission while the use of two frequencies allows full-duplex (simultaneous two-way) transmission just like a wired telephone. A choice of several frequencies is available so that neighbors can use different frequencies to prevent interference and eavesdropping. Also, signaling is done by guard tones in sequences which are selectable and unlikely to be duplicated in the neighborhood.

MOBILE TELEPHONES

Mobile telephones may be thought of as cordless phones with elaborate portable and base units. High-power transmitters and elevated antennas that provide the radio carrier link over an area within 20 to 30 miles from the base station antenna, as well as the multiplexing, detecting, sorting and selecting features required to simultaneously service 60 subscribers per base station, are the major differences between cordless telephones and mobile telephones.

Base Unit

Figure 10-2 shows a mobile telephone system. The base station can transmit and receive on several different frequencies simultaneously to provide several individual channels for use at the same time. The radio base station transmitter output power is typically 200-250 watts and the radiated power can be as high as 500 watts if the transmitting antenna gain is included. It covers a circular area of up to 30 miles in radius for clear reliable communications, but transmitters with the same frequency are not spaced closer than about 60 to 100 miles because of the noise interference levels.

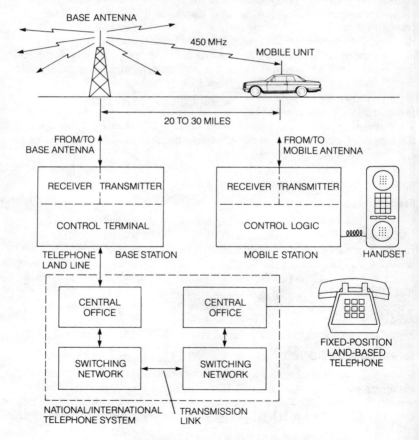

Figure 10-2. *Mobile Telephone System*

The receiver contains filters, high-gain amplifiers, and demodulators to provide a usable voice signal to the telephone line. The control terminal contains the necessary detector and timing and logic circuits to control the transmission link between the base unit and the mobile units. As a result, telephone calls are coupled to and from the standard telephone system just like calls that are carried completely over wired facilities. The control terminal has the necessary interface circuits so that a call initiated at a mobile unit is interconnected through the national or international telephone system to the called party just as any other telephone call.

The national and international telephone system facilities are owned by the respective telephone companies. The base units and mobile units may be owned by the telephone company or by a separate company called a radio common carrier (RCC). When the mobile system is run by a RCC, the RCC is charged by the telephone company for the use of the standard telephone system just like any other customer. This cost is then included in the charge by the RCC to the eventual user of the mobile units.

To subscribe to mobile telephone service, a user has only to apply and be accepted by the RCC or the telephone company operating the system. When the application is accepted, the user can lease or purchase the mobile equipment.

Mobile Unit

The mobile unit in the user's vehicle consists of a receiver containing amplifiers, a mixer and a demodulator; a transmitter containing a modulator, carrier oscillators and amplifiers; the necessary control logic; a control unit with microphone, speaker, keypad and switches; antennas and the interconnecting cables. The control unit performs all of the functions associated with normal telephone use. A modern control head with automatic functions is illustrated in *Figure 10-3*.

The mobile telephone user with automatic control places and receives calls in the same manner as with an ordinary telephone. When the handset is lifted to place a call, the radio unit automatically selects an available channel. If no channel is available, the busy light comes on. If a channel is found, the user hears the normal dial tone from the telephone system, and can then dial the number and proceed as if the telephone were direct wired. An incoming call to the mobile unit is signaled by a ringing tone and is answered simply by lifting the handset and talking. Thus, the automatic mobile telephone is as easily used as a home telephone. The mobile telephone combines the mobility of the radio link and the world-wide switched network of the existing telephone system to provide a communication link to any other telephone in the world.

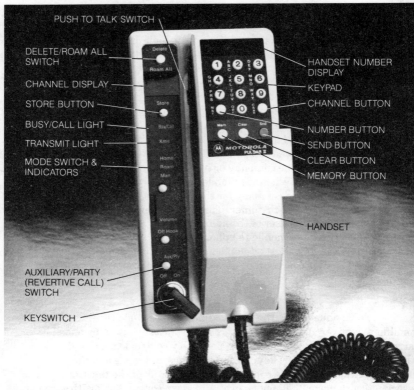

Figure 10-3. Control Head for Mobile Telephone
(Courtesy of Motorola, Inc. Pulsar II is a trademark of Motorola, Inc.)

Home Area and Roaming

As previously stated, the mobile system is designed for optimum use within a 20-30 mile radius of the base antenna. This is called the subscriber's home area and a subscriber usually would remain in the home area. However, if the subscriber moves out of the home area into another area, the subscriber is referred to as a roamer and a different mode of operation applies.

Each mobile telephone has a unique telephone number which includes the home area's base station identification. When someone calls the mobile unit, the calling party is connected first to the transmitter serving the subscriber's home area. As long as the subscriber is within radio range of that system, all is well; otherwise, the base station won't get an answer from the mobile unit and the caller will get a no-answer signal. If the subscriber roams outside the home area, he/she can still be reached if a similar mobile telephone system exists in that area, provided proper advance arrangements have been made.

Calls to roamers are usually placed by calling a special number for the mobile service operator who knows the roamer's location. The operator manually patches the call through to the base station serving the area of the roamer's location. Some systems cannot handle roamers due to overload of their channels, and some systems do not allow roamers.

Detailed Operation

Different signaling techniques must be used in a mobile telephone system than in a wired facility. Since there are no wires connecting the telephone to the network, both speech and signaling must be transmitted via radio. This is accomplished through the use of special tones rather than applying a voltage level or detecting a current. The tones are selected so as not to be mistaken for other signaling tones, such as DTMF. The proper tone transmitted to the mobile unit will, for example, ring the mobile telephone to indicate an incoming call just as with a standard telephone. A different tone is used to indicate off-hook, busy, etc.

The Improved Mobile Telephone System (IMTS) uses in-band signaling tones from 1,300 Hz to 2,200 Hz. The older Mobile Telephone System (MTS) had in-band signaling tones in the 600 Hz to 1,500 Hz range. Some systems use 2,805 Hz in manual operation.

Incoming Call

To gain a better understanding of the system operation, let's trace an incoming call from a wire facility subscriber through the base unit to a mobile unit. The base station controls all activity on all channels and can transmit on any idle channel. Regardless of how many channels are idle, it selects only one and places a 2,000 Hz idle tone on it as shown in *Figure 10-4*. All on-hook mobile units that are turned on automatically search for the idle tone and lock on the idle channel because this is the channel over which the next call in either direction will be completed. After locking onto the idle channel, all on-hook mobile units "listen" for their number on that channel. When an idle channel becomes busy for a call in either direction, the base station control terminal selects another unused channel and marks it with the idle tone. All on-hook mobile units then move to the new idle channel. This process is repeated each time a new call is initiated as long as unused channels are available.

After the person calling the mobile subscriber dials the mobile unit's telephone number, the call is processed through the switched telephone network as in a normal landline call. Following the sequence in *Figure 10-4*: When the call reaches the control terminal, the terminal seizes the idle channel, and indicates seizure by removing the idle tone from that channel and applying the 1,800 Hz seize-tone. The seize-tone prevents mobile units from seizing the channel to originate a call. The control terminal then out-pulses the mobile unit's number over the base station transmitter at ten pulses per second, with idle-tone representing a mark (which corresponds to the make interval in dc pulsing) and seize-tone representing a space (corresponding to the break interval).

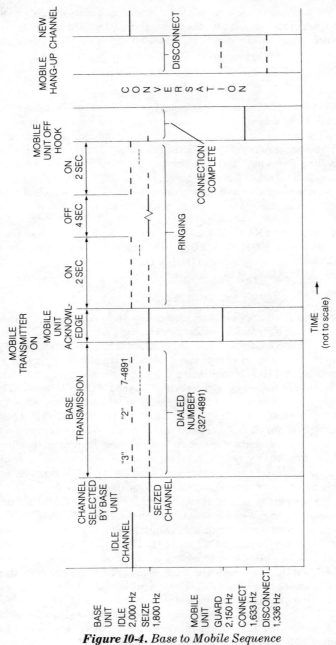

Figure 10-4. *Base to Mobile Sequence*

Each on-hook mobile unit receiving the number transmission compares the received number to its unit number. As soon as a digit mismatch is detected, the mobile unit abandons that channel and searches for the new idle channel. Thus, upon completion of the number transmission, all mobile units except the one called will have abandoned the seized channel and will be monitoring the new idle channel.

When the mobile unit receives its correct seven-digit address, the mobile supervisory unit turns on the mobile transmitter and sends the acknowledgement signal, using the 2,150 Hz guard-tone, back to the control terminal. If this acknowledgement is not received by the control terminal within three seconds after out-pulsing the address, the seize tone is removed and the call abandoned. However, upon receipt of the mobile acknowledgement signal, the terminal sends standard repetitive ringing at a cycle of two seconds on, four seconds off, using idle and seize tones as before. If the mobile does not answer within 45 seconds, ringing is discontinued and the call abandoned.

When the mobile subscriber goes off-hook to answer, the mobile supervisory unit sends a burst of connect tone (1,633 Hz) as an answer signal. Upon receipt of the answer signal, the control terminal stops the ringing and establishes a talking path between the calling circuit and the radio channel. When the subscriber hangs up at the end of the call, the mobile supervisory unit sends a disconnect signal—alternating the disconnect tone (1,336 Hz) and the guard tone. The mobile supervisory unit then turns off the mobile transmitter and begins searching for the marked idle channel.

Outgoing Call

The sequence for a call originated by a mobile subscriber is illustrated by *Figure 10-5*. When the subscriber goes off-hook to place the call, the mobile unit must be locked on the marked-idle channel. If not, the handset will be inoperative and the busy lamp on the control unit will light, indicating to the subscriber that no channel is available. If the mobile unit is locked on the marked-idle channel, the mobile supervisory unit will turn on the mobile transmitter to initiate the acknowledgement or handshake sequence. The identification section of *Figure 10-5* is where the mobile unit transmits its own number so the control terminal can identify it as a subscriber and can charge the call to the number. The pulses of guard tone mixed in with the number pulses are for parity checking. The remaining functions of *Figure 10-5* are similar to those of *Figure 10-4*.

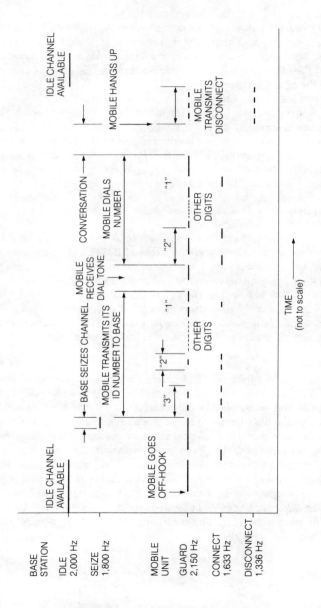

Figure 10-5. *Mobile to Base Sequence*

CELLULAR MOBILE TELEPHONE SERVICE

Mobile telephone service always has been a scarce luxury. Subscribers pay from ten to twenty times more for mobile service than for residential telephone service, yet most urban telephone carriers and RCCs have long waiting lists for mobile telephones. In Chicago, for example, only 2,000 mobile users can be accommodated, yet at least ten times that many desire service at present rates. The reason is that there simply are not enough channels to handle the demand, and the few dozen available are spread over several bands and divided among different types of carriers. The solution is not simply to assign new frequencies and build more transmitters because the spectrum space for new frequencies is simply not available; besides, this would not eliminate the restrictions on roamers. Clearly, an entirely new approach to mobile telephony was needed. The cellular concept, also called the Advanced Mobile Phone Service (AMPS)[1] is a method to provide high quality mobile service for more subscribers at an affordable cost and to provide more freedom for roamers.

Basic Concept

The basic concept of the AMPS cellular system is to reduce the area covered by the transmitter by reducing the power of transmission. In this way, concentrated areas of population can have more transmitting stations, and thus more channels, because each transmitter handles a given number of conversations. In addition, because transmitters cover less area, the same frequency can be re-used in a common geographical area.

System Structure

The basic system arrangement is shown in *Figure 10-6*. The service area is divided into regions called cells, each of which has equipment to switch, transmit, and receive calls to/from any mobile unit located in the cell. Each cell transmitter and receiver operates on a given channel. Each channel is used for many simultaneous conversations in cells which are not adjacent to one another, but are far enough apart to avoid excessive interference. Thus a system with a relatively small number of subscribers can use large cells, and as demand grows, the cells are divided into smaller ones.

The MTSO

The cell sites are interconnected and controlled by a central Mobile Telecommunications Switching Office (MTSO), which is basically a telephone switching office as far as hardware is concerned, but as shown in *Figure 10-6*, it uses a substantial amount of additional digital equipment programmed for cellular control. It not only connects the system to the telephone network, but also records call information for billing purposes. The MTSO is linked to the cell sites by a group of voice trunks for conversations, together with one or more data links for signaling and control. The MTSO controls not only the cell sites via radio commands, but also many functions of the mobile units.

[1]Advanced Mobile Phone Service (AMPS) is a registered service mark of AT&T Co.

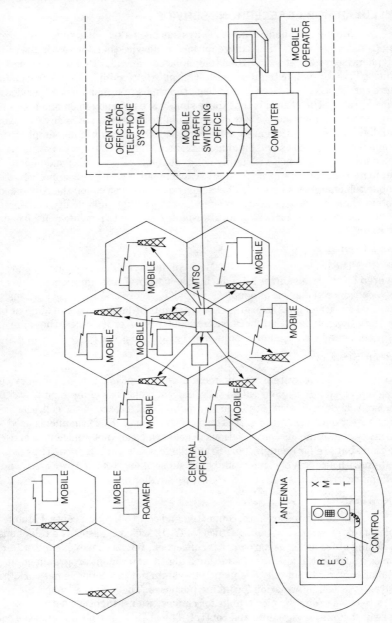

Figure 10-6. Cellular Network

Mobile Units

The mobile units consist of a control unit, a transceiver, and appropriate antennas. The transceiver contains circuits that can tune to any of the 666 FM channels in the 800 MHz range assigned to the cellular system. Each cell site has at least one setup channel dedicated for signaling between the cell and its mobile units. The remaining channels are used for conversation.

Each mobile unit is assigned a 10-digit number, identical in form to any other telephone number. Callers to the mobile unit will dial the local or long-distance number for the desired mobile unit. The mobile user will dial seven or ten digits with a 0 or a 1 prefix, where applicable, as if calling from a fixed telephone.

Whenever a mobile unit is turned on but not in use, the mobile control unit monitors the data being transmitted on a setup channel selected from among the several standard setup frequencies on the basis of signal strength. If signal strength becomes marginal as the mobile unit approaches a cell boundary, the mobile control finds a setup channel with a stronger signal.

Calling a Mobile Unit

A mobile unit is called by transmitting its number over the setup channel. When the mobile recognizes its number, it quickly seizes the strongest setup channel and transmits an acknowledgement response. The cell site then uses the seized setup channel to transmit the voice-channel assignment to the mobile. The mobile and cell site switch to the voice-channel radio frequency and the voice channel is used for ringing, off-hook, and subsequent conversation. The sequence is similar when the mobile user originates the call. It begins with the mobile control unit seizing a setup channel when the mobile unit goes off-hook. Then the voice channel selection, signaling and conversation occur in the same way.

During the call, the system at the serving cell site examines the signal strength once every few seconds. If the signal level becomes too low, the MTSO looks for a cell site closer to the mobile unit to handle the call, based on the location and direction of travel information from the serving cell site. The actual handoff from one cell to the next occurs so rapidly that the user cannot tell it has occurred.

Roamers

The system is designed to make handling of roamers automatic; indeed, this is the principal goal of the cellular approach. Locating and handoff are concepts that come directly from the use of small cells. "Locating" in this sense is not the determination of precise geographic location—although that is obviously a factor; rather, it is the process of determining whether a moving active user should continue to be served by his current channel and transmitter, or "handed off" to either another channel, cell, or both. The decision is made automatically by a computer, based on signal quality and potential interference, and involves sampling the signal from the mobile unit.

With the cellular system, a subscriber could make a call from his car while driving in the countryside toward a city, continue through the city's downtown, and not hang up until well beyond the city on the other side. During the entire time, the transmission would be clear with no dead spots. More importantly, the switching of transmitters and frequencies during the conversation would be entirely automatic, with no interruptions and *no action required by the user or an operator*.

Wherever there is a system to serve it, a roaming unit will be able to obtain completely automatic service; however, a call from a land telephone to a mobile unit which has roamed to another metropolitan area presents additional problems. While it would be technically possible for the system to determine automatically where the mobile unit is, and to connect it automatically to the land party, there are two reasons for not doing so. First, the caller will expect to pay only a local charge if a local number is dialed. Second, the mobile user may not want to be identified to be at a particular location automatically by the system without an approval. Therefore, the system will complete the connection only if the extra charge is agreed to, and when possible to do so without unauthorized disclosure of the service area to which the mobile unit has roamed.

Unique Features

There are two essential elements of the cellular concept which are unique; frequency reuse and cell splitting.

Frequency reuse means using the same frequency or channel simultaneously for different conversations, in the same general geographic area. The idea of having more than one transmission on a given frequency is not new; it is done in virtually all radio services. What is unique to cellular is the closeness of the users; two users of the same frequency may be only a few dozen miles apart, rather than hundreds of miles. This is done by using relatively low-power transmitters on multiple sites, rather than a single high-power transmitter. Each transmitter covers only its own cell, and cells sufficiently far apart may be using the same frequency.

Cell splitting is based on the notion that cell sizes are not fixed, and may vary in the same area or over time. The principle is shown in *Figure 10-7*. Initially, all the cells in an area may be relatively large as shown in *Figure 10-7a*. When the average number of users in some cells becomes too large to be handled with proper service quality, the overloaded cells are split into smaller cells by adding more transmitters, as shown in *Figure 10-7b*. The same MTSO can continue to serve all of the cell sites, but expansion of its computer and switching facilities probably will be required.

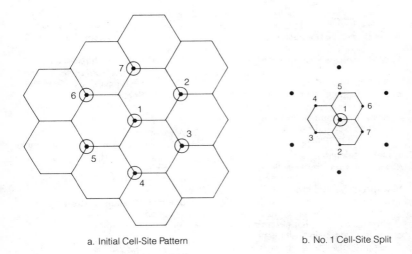

a. Initial Cell-Site Pattern   b. No. 1 Cell-Site Split

Figure 10-7. *Cell Splitting*
(Reprinted with permission from Bell System Technical Journal, *Copyright© 1979,*
American Telephone and Telegraph Co.)

WHAT HAVE WE LEARNED?

1. Wireless telephones consist of a base unit and a mobile unit that are connected by a radio link instead of wires.
2. Wireless telephones may be cordless phones for the home or small business or mobile radio telephones used in commerce and industry.
3. Most cordless telephones use frequency modulation to carry signaling and voice communications.
4. Many mobile telephones have a high-power transmitter at the base station which covers an area with a radius of 20-30 miles.
5. Specific sequences and tone signaling must be followed to complete the radio link correctly between a mobile telephone base and mobile unit.
6. The cellular wireless telephone system was designed to allow many more subscribers to use wireless telephones.
7. A cellular system handles the switching of cell sites and signal channels automatically as a roamer moves through the grid of cells.

Quiz for Chapter 10

1. As used in this book, the term "wireless telephone" means:
  a. trans-oceanic telephone service via radio.
  b. a telephone where the wiring has been replaced by solid-state integrated circuits.
  c. any telephone device or system which uses a radio link to replace a wired link somewhere between the handset and the public switched network.

2. A mobile telephone is:
  a. another name for a cordless telephone.
  b. a special form of radio telephone that connects with the public switched telephone network through the telephone company.
  c. cordless extension telephone designed for use in automobiles.

3. Mobile telephone users originate and receive calls:
  a. by going through the regular telephone operator.
  b. automatically, provided they are in their home area and a channel is available.
  c. by using two different radio channels, one for transmission and one for reception.

4. Cellular mobile telephone service is:
  a. the concept of using many low-power transmitters with computer control, rather than a few high-power transmitters.
  b. another name for IMTS.
  c. the concept of improving frequency utilization by restricting users to a single or home area called a cell.

5. A cordless telephone:
  a. uses fiber-optic techniques to replace the traditional cord.
  b. is a portable extension telephone using short-distance radio links between the handset and the base unit.
  c. uses radio links between the base unit and the telephone company lines.

6. Antennas for a cordless telephone:
  a. are usually placed on the roof, sometimes combined with the TV antenna.
  b. consist of two whips, one for transmission from the base, the other for transmission from the portable.
  c. consist of the building wiring, whips, and loopsticks.

7. A roamer is:
  a. a user of a cordless telephone who walks around the building while talking.
  b. a term that applies only to cellular systems.
  c. a mobile telephone user who is outside of the home area.

8. When a cellular mobile telephone moves from one cell to another while a call is in progress,
  a. the call must be terminated, then initiated from the new cell.
  b. the system automatically transfers the call to another transmitter, and possibly to another channel.
  c. the user must make arrangements with the telephone company to have calls forwarded.

9. An idle channel in the IMTS system is:
  a. the condition of each mobile unit's dedicated channel when it is not being used.
  b. one unused channel which all mobile units monitor for their address.
  c. a special channel dedicated for signaling only, which is often idle.

10. Cell splitting is:
  a. the process of dividing a large mobile telephone cell into smaller cells when usage requires.
  b. the idea of two mobile units using the same frequency at different locations in the same cell.
  c. a process similar to splitting hairs, often used by people who make up multiple-choice quizzes.

Glossary

A/D Converter: A circuit that converts signals from analog form to digital form.

Address: The number dialed by a calling party which identifies the party called. Also a location or destination in a computer program.

Aliasing: The occurrence of spurious frequencies in the output of a PCM system that were not present in the input, due to foldover of higher frequencies.

AM (Amplitude Modulation): A technique for sending information as patterns of amplitude variations of a carrier sinusoid.

Amplifier: An electronic device used to increase signal power or amplitude.

Analog: Information represented by continuous and smoothly varying signal amplitude or frequency over a certain range, such as in human speech or music.

Asynchronous: Refers to circuitry and operations without common timing (clock) signals.

Attenuation: The decrease in power that occurs when any signal is transmitted.

Audio Frequency: Frequencies detectable by the human ear, usually between 20 and 15,000 Hz.

Bandwidth: The range of signal frequencies that a circuit or network will respond to or pass.

Base Unit: The transmitter (antenna and equipment), in a fixed location, and usually having higher power than the mobile units.

Binary Code: A pattern of binary digits (0 and 1) used to represent information such as instructions or numbers.

Bipolar: Having both positive and negative polarity.

Bit: An acronym for binary digit; the smallest piece of binary information; a specification of one of two possible alternatives.

BORSCHT: An acronym for the functions that must be performed in the central office when digital voice transmission occurs: Battery, Overvoltage, Ringing, Supervision, Coding, Hybrid, and Test.

Byte: A group of 8 bits treated as a unit. Often equivalent to one alphabetic or numeric character.

Cable: An assembly of one or more conductors insulated from each other and from the outside by a protective sheath.

Cell: In cellular mobile telephony, the geographic area served by one transmitter. Subscribers may move from cell to cell.

Central Office (CO): The switching equipment that provides local exchange telephone service for a given geographical area, designated by the first three digits (NNX or NXX) of the telephone number.

Channel: An electronic communications path, usually of 4,000 Hz (voice) bandwidth.

Circuit: An interconnected group of electronic devices, or the path connecting two or more communications terminals.

Common Battery: A system of supplying direct current for the telephone set from the central office.

Compander: An acronym for COMpressor-ExPANDER, a circuit that compresses the dynamic range of an input signal, and expands it back to almost original form on the output.

Crossbar Switch: An electromechanical switching machine utilizing a relay mechanism with horizontal and vertical input lines (usually 10 by 20), using a contact matrix to connect any vertical to any horizontal.

Crosspoint: The element that actually performs the switching function in a telephone system. May be mechanical using metal contacts, or solid-state using integrated circuits.

Crosstalk: Undesired voice-band energy transfer from one circuit to another (usually adjacent.)

Current: The flow of electrical charge, measured in amperes.

Cut-Off Frequency: The frequency above which or below which signals are blocked by a circuit or network.

D/A Converter: A circuit that converts signals from digital form to analog form.

Data: In telephone systems, any information other than human speech.

Data Set: Telephone company term for modem.

Decibel (dB): A unit of measure of relative power or voltage, in terms of the ratio of two values. One dB = 10 log P1/P2, where P1 and P2 are the power levels in watts.

Decoder: Any device which modifies transmitted information to a form which can be understood by the receiver.

Demodulation: The process of extracting transmitted information from a carrier signal.

Demultiplexer: A circuit that distributes an input signal to a selected output line (with more than one output line available).

Digital: Information in discrete or quantized form; not continuous.

Distortion: Any difference between the transmitted and received waveforms of the same signal.

DTMF (Dual-Tone-Multi-Frequency): Use of two simultaneous voice-band tones for dialing.

Electromagnetic Spectrum: The entire available range of sinusoidal electrical signal frequencies.

Encoder: Any device which modifies information into the desired pattern or form for a specific method of transmission.

ESS (Electronic Switching System): A telephone switching machine using electronics, often combined with electromechanical crosspoints, and usually with a stored-program computer as the control element.

Exchange Area: The territory within which telephone service is provided without extra charge. Also called the Local Calling Area.

FCC (Federal Communications Commission): A government agency that regulates and monitors the domestic use of the electromagnetic spectrum for communications.

Fiber Optics: The process of transmitting infrared and visible light frequencies through a low-loss glass fiber with a transmitting laser or LED.

FM (Frequency Modulation): A technique for sending information as patterns of frequency variations of a carrier signal.

Frequency: The rate in hertz (cycles per second) at which a signal pattern is repeated.

FSK (Frequency-Shift Keying): A method of transmitting digital information that utilizes two tones; one representing a one level, the other a zero level.

Ground: An electrical connection to the earth or to a common conductor which is connected to the earth at some point.

Ground Start: A method of signaling between two machines where one machine grounds one side of the line and the other machine detects the presence of the ground.

Half-Duplex: A circuit that carries information in both directions, but only in one direction at a time.

Home Area: The geographic area in which a mobile telephone subscriber is normally located.

Hybrid: In telephony, a circuit that divides a single transmission channel into two, one for each direction; or conversely, combines two channels into one.

Instruction Code: Digital information that represents an instruction to be performed by a computer.

Integrated Circuit: A circuit whose connections and components are fabricated into one integrated structure on a certain material such as silicon.

Lineside: Refers to the portion of the central office that connects to the local loop.

Local Loop: The voice-band channel connecting the subscriber to the central office.

Loop Start: The usual method of signaling an off-hook or line seizure, where one end closes the loop and the resulting current flow is detected by the switch at the other end.

Loss: Attenuation of a signal from any cause.

Message Telephone Service (MTS): The official name for long distance or toll service.

Microwaves: All frequencies in the electromagnetic spectrum above one billion hertz (1 gigahertz).

Mobile Unit: The part of the mobile telephone system that is not fixed in its location; hence, is not attached to the telephone network by wires.

Modulation: The systematic changing of the properties of an electronic wave, using a second signal, to convey the information contained in the second signal.

Multiplexing: The division of a transmission facility into two or more channels.

Off-Hook: The condition that indicates the active state of a customer telephone circuit. The opposite condition is On-Hook.

Oscillator: An electronic device used to produce repeating signals of a given frequency and amplitude.

PABX or PBX: A private (automatic) branch telephone exchange system providing telephone switching in an office or building.

Parallel Data: The transfer of data simultaneously over two or more wires or transmission links.

Parity: A bit that indicates whether the number of "ones" in a bit string is odd or even.

PCM (Pulse Code Modulation): A communication systems technique of coding signals with binary codes to carry the information.

Period: The time between successive similar points of a repetitive signal.

Phase: The time or angle that a signal is delayed with respect to some reference position.

Portable Unit: The portion of a cordless telephone that is not electrically attached to the network by wires. Consists of a transmitter, receiver, and perhaps a keypad.

POTS: "Plain Old Telephone Service", single line residential rotary dial service.

Program: The sequence of instructions stored in the computer memory.

RCC: Radio Common Carrier; a company that provides mobile telephone service, but is not a telephone company.

Receiver: The person or device to which information is sent over a communication link.

Register: A series of identical circuits placed side-by-side that are able to store digital information.

Ring: The alerting signal to the subscriber or terminal equipment; the name for one conductor of a wire pair, designated by R.

Roamer: A mobile telephone subscriber using the system outside of his/her home area.

Serial Data: The transfer of data over a single wire in a sequential pattern.

Sidetone: That portion of the talker's voice which is fed back to his receiver.

Simplex: A circuit which can carry information in only one direction; for example, broadcasting.

SLIC (Subscriber Line Interface Circuit): In digital transmission of voice, the circuit which performs some or all of the interface functions at the central office. See BORSCHT.

State: A condition of an electronic device, especially a computer, which is maintained until an internal or external occurrence causes change.

Step-by-Step (SxS) System: An electromechanical telephone switching system in which the switches are controlled directly by digits dialed by the calling party.

Subscriber: The telephone customer.

Subscriber Loop: Another term for local loop.

Synchronous: Events that are controlled by or referred to a common clock.

TDM (Time Division Multiplexing): A communication system technique that separates information from channel inputs and places them on a carrier in specific positions in time.

Tip: One conductor of a wire pair, designed by T; usually the more positive of the two.

Toll Center: A major telephone distribution center that distributes calls from one major metropolitan area to another.

Transmission: Passing information, using electromagnetic energy, from one point to another.

Transmission Link: The path over which information flows from sender to receiver.

Transmitter: The person or device that is sending information over a communication link.

Trunk: A transmission channel connecting two switching machines.

Trunkside: The portion of the central office that connects to trunks going to other switching offices.

Voice-Grade Line: A local loop, or trunk, having a bandpass of approximately 300 to 3,000 Hz.

Voltage: A measure of the electrical force that causes current flow in a circuit.

Wideband Circuit: A transmission facility having a bandwidth greater than that of a voice-grade line.

Index

Answers to Quizzes

Chapter 1
1. b
2. a
3. d
4. d
5. b
6. d
7. a
8. c
9. b
10. d
11. d
12. b

Chapter 2
1. b
2. a
3. b
4. b
5. d
6. c
7. a
8. d
9. b
10. d

Chapter 3
1. c
2. b
3. c
4. a
5. c
6. d
7. c
8. b
9. a
10. b

Chapter 4
1. c
2. d
3. a
4. c
5. a
6. d
7. b
8. d
9. c
10. b

Chapter 5
1. b
2. c
3. b
4. a
5. b
6. a
7. b
8. e
9. b
10. c

Chapter 6
1. b
2. d
3. d
4. b
5. c
6. c
7. d
8. d
9. b
10. a
11. c
12. b
13. b
14. a
15. b

Chapter 7
1. c
2. c
3. b
4. a
5. b
6. c
7. c
8. b
9. a
10. b

Chapter 8
1. e
2. d
3. e
4. e
5. d
6. a
7. d
8. a
9. b
10. d

Chapter 9
1. b
2. f
3. c
4. c
5. d
6. a
7. d
8. a
9. e
10. d

Chapter 10
1. c
2. b
3. b
4. a
5. b
6. c
7. c
8. b
9. b
10. a